计算机科学与技术专业实践系列教材

ASP.NET
网站设计实例教程

邓 芳　毕忠东　刘启明　主　编
牟德昆　刘晓梅　程立倩　副主编

清华大学出版社
北京

内 容 简 介

本书深入浅出、循序渐进地讲授如何使用 ASP.NET 进行系统开发，内容包括 ASP.NET 的基础理论、基本控件的使用、内置对象的使用、ADO.NET 的基础、数据库基础、ASP.NET 操作数据库的方式等。

全书共分 9 章：第 1 章～第 7 章主要讲授 ASP.NET 的基础理论，最后两章是实际案例，在了解了基本的模块开发后，最后两章进行了较大、较完善的系统开发的详细讲解。具体内容包括用户注册模块、登录模块、显示新闻模块、分页显示模块等方面，全书提供了大量应用实例。

本书既可作为计算机专业的专业课程教材，又可以作为网页制作爱好者、ASP.NET 应用程序初学者、.NET 应用开发入门者的指导书。

本书封面贴有清华大学出版社防伪标签，无标签者不得销售。
版权所有，侵权必究。侵权举报电话：010-62782989　13701121933

图书在版编目(CIP)数据

ASP.NET 网站设计实例教程/邓芳，毕忠东，刘启明主编. --北京：清华大学出版社，2015(2019.8重印)
计算机科学与技术专业实践系列教材
ISBN 978-7-302-40057-8

Ⅰ. ①A…　Ⅱ. ①邓…②毕…③刘…　Ⅲ. ①网页制作工具－程序设计－高等学校－教材
Ⅳ. ①TP393.092

中国版本图书馆 CIP 数据核字(2015)第 089392 号

责任编辑：白立军
封面设计：傅瑞学
责任校对：时翠兰
责任印制：刘海龙

出版发行：清华大学出版社
　　网　　址：http://www.tup.com.cn, http://www.wqbook.com
　　地　　址：北京清华大学学研大厦 A 座　　邮　编：100084
　　社 总 机：010-62770175　　邮　购：010-62786544
　　投稿与读者服务：010-62776969, c-service@tup.tsinghua.edu.cn
　　质量反馈：010-62772015, zhiliang@tup.tsinghua.edu.cn
　　课件下载：http://www.tup.com.cn, 010-62795954

印 装 者：北京九州迅驰传媒文化有限公司
经　　销：全国新华书店
开　　本：185mm×260mm　　印　张：11　　字　数：264 千字
版　　次：2015 年 6 月第 1 版　　印　次：2019 年 8 月第 3 次印刷
定　　价：25.00 元

产品编号：063293-01

序

当今世界,以计算机技术、通信技术和控制技术为代表的3C技术正迅猛发展,而以Internet为代表的全球范围内信息基础设施的建设成就,标志着人类社会已进入信息时代。应用型人才培养是社会发展和高等教育发展的必然要求,经济社会发展迫切需要高等学校培养出在知识、能力、素质等诸方面都适应社会需要的不同层次的应用型人才,以满足信息化社会建设的需求。实施高等教育名校建设工程,对大力发展高等教育,指导高等教育特色发展,全面提高教育质量,增强高等教育的竞争力和服务经济社会能力具有重大意义。

罗伯特·加涅是美国教育心理学家,加涅将认知学习理论应用于教学过程,加涅理论中的引出作业和提供反馈是一种教学策略。从"应用型"人才培养的角度来说,学生的实践能力提升是一个重要问题,需要学校和教师采取一些有效手段来增强学生的实践能力,树立以学生为本的观念,尊重学生的个性特点,因材施教,增加学生对于课程、专业的选择空间。

本套教材(指以刘启明教授主编的教材)是我们多年来进行"应用型人才培养教学内容、课程体系改革"的综合成果。实施教学方式、教学内容、考核机制的全面改革,在培养学生的信息能力和信息素养方面具有先导作用。我们提出的课程内容设置方案,目的是推进人文与自然的有机融合,适应学生能力、兴趣、个性、人格全面发展的需要,强化学生的实践能力和创新能力培养,为计算机课程教学内容、课程体系改革,设计了一个全新的框架。

本套教材以应用型、技能型人才培养为目标,以重点专业建设为平台,围绕着教育教学改革、创新人才培养、提高人才培养质量的教育发展理念展开。每部教材都是由应用型名校计算机专业课教师或者计算机实验教学示范中心专业教师编写完成的。

在本套教材的编写过程中,我们得到许多专家的精心指点和热情帮助。教育部计算机科学与技术教学指导委员会先后三次在我校召开计算机课程教学研讨会,清华大学、北京大学、中国人民大学、复旦大学、浙江大学、南京大学、中国科学技术大学等近百所高校的老师参加。专家学者对本套教材的编写提出了很多宝贵意见。

本套教材的出版得到清华大学出版社的大力支持,正是他们精益求精的工作,才使这一系统工程得以顺利完成,并得到高度评价,在此表示衷心感谢。

刘启明

2015年1月

前　　言

随着互联网的不断发展和平台的多样化，越来越多的Web开发技术呈现在用户面前，也是由于互联网的不断发展，越来越多的普通用户进入了互联网的范围开始了网络生活。而人们使用的互联网的应用，通常情况是通过一些Web编程语言进行实现的，这些语言包括ASP.NET、JSP和PHP等。Web开发技术的不断完善，更多丰富的应用程序也随之诞生，ASP.NET使用.NET平台进行Web应用程序开发有着先天的优势，开发人员能够快速地使用ASP.NET提供的控件进行复杂的应用程序开发。

本书以高素质应用型、技能型人才培养为目标，以重点专业建设为平台，围绕着不断深化教育教学改革、创新人才培养模式、提高人才培养质量、实现高等教育又好又快发展的教育理念。

本书全面介绍ASP.NET技术，附带大量的实例及详细的注释，方便初学者深入学习。本书的主要特点是通过将一些固定的知识进行分类讲解，举一反三，通过简单的实例讲解复杂的理论知识，让学生能够通过案例来对所学内容进行深入理解。再一个特点就是本教程不是面面俱到，不是把所有ASP.NET的知识都罗列出来，而是仅仅讲述人们在做项目中经常使用到的相应知识，对于不常用的知识，让学生学会如何查找即可。

本书由邓芳、毕忠东、刘启明担任主编，牟德昆、刘晓梅、程立倩担任副主编，王作鹏教授、崔玉礼教授审阅了书稿。

本书既可作为计算机专业的专业课程教材，也可以作为网页制作爱好者、ASP.NET应用程序初学者、.NET应用开发入门者的指导书。

在本教材的编写、出版过程中，我们得到许多专家的精心指点和热情帮助。清华大学出版社给予了大力支持，正是他们精益求精的工作，才使这一系统工程得以顺利完成，并得到高度评价，在此表示衷心感谢。

限于作者水平，书中难免存在谬误之处，恳请广大专家、读者提出宝贵意见。

作　者

2015年2月

目 录

第1章 ASP.NET 初识 ... 1
1.1 ASP.NET 简介 ... 1
1.1.1 ASP.NET 发展历程 ... 1
1.1.2 ASP.NET 能做哪些事情 ... 1
1.1.3 ASP.NET 的运行原理 ... 2
1.2 制作一个 ASP.NET 网站 ... 2
1.2.1 创建 ASP.NET 网站 ... 2
1.2.2 写入代码 ... 3
1.2.3 运行应用程序 ... 4
1.3 小结 ... 5
课后思考问题 ... 5

第2章 ASP.NET 内置对象 ... 6
2.1 Response 对象 ... 6
2.2 Request 对象 ... 7
2.2.1 QueryString 集合 ... 7
2.2.2 Form 集合 ... 9
2.3 Server 对象 ... 10
2.4 小结 ... 11
课后思考问题 ... 11

第3章 标准控件 ... 12
3.1 Label 控件 ... 12
3.1.1 Label 控件的属性 ... 12
3.1.2 Label 控件的实际使用 ... 12
3.2 TextBox 控件 ... 13
3.2.1 TextBox 控件基本属性 ... 13
3.2.2 TextBox 控件的实际使用 ... 14
3.3 DropDownList 控件 ... 15
3.3.1 DropDownList 控件添加列表项 ... 15
3.3.2 获取 DropDownList 控件列表项的值 ... 17
3.3.3 DropDownList 控件的实际应用 ... 17
3.4 ListBox 控件 ... 18
3.4.1 ListBox 控件添加列表项 ... 18
3.4.2 ListBox 控件实际使用 ... 18

3.4.3　ListBox 控件实现多项选择 ……………………………………………… 19
3.5　CheckBox 控件 ………………………………………………………………… 20
　　3.5.1　CheckBox 控件基本属性 …………………………………………………… 20
　　3.5.2　CheckBoxList 控件 …………………………………………………………… 22
3.6　RadioButton 控件 ……………………………………………………………… 23
　　3.6.1　RadioButton 控件基本属性 ………………………………………………… 23
　　3.6.2　RadioButtonList 控件 ………………………………………………………… 24
3.7　Image 控件 ……………………………………………………………………… 25
3.8　按钮控件 ………………………………………………………………………… 26
　　3.8.1　按钮控件的属性 ……………………………………………………………… 26
　　3.8.2　按钮控件的 Click 事件 ……………………………………………………… 26
3.9　HyperLink 控件 ………………………………………………………………… 27
3.10　FileUpload 控件 ………………………………………………………………… 28
3.11　Calendar 控件 …………………………………………………………………… 30
3.12　小结 ……………………………………………………………………………… 31
课后思考问题 …………………………………………………………………………… 31

第 4 章　验证控件的使用 ……………………………………………………………… 32
4.1　验证控件概述 …………………………………………………………………… 32
4.2　RequiredFieldValidator 控件 ……………………………………………………… 32
　　4.2.1　RequiredFieldValidator 控件的基本属性 …………………………………… 32
　　4.2.2　RequiredFieldValidator 控件的实际使用 …………………………………… 33
4.3　RangeValidator 控件 ……………………………………………………………… 34
　　4.3.1　RangeValidator 控件的基本属性 …………………………………………… 34
　　4.3.2　RangeValidator 控件的实际使用 …………………………………………… 34
4.4　CompareValidator 控件 …………………………………………………………… 35
　　4.4.1　CompareValidator 控件的基本属性 ………………………………………… 35
　　4.4.2　CompareValidator 控件在实际中的使用 …………………………………… 35
4.5　RegularExpressionValidator 控件 ………………………………………………… 36
　　4.5.1　RegularExpressionValidator 控件的基本属性 ……………………………… 36
　　4.5.2　RegularExpressionValidator 控件在实际中的使用 ………………………… 36
4.6　CustomValidator 控件 …………………………………………………………… 37
4.7　小结 ……………………………………………………………………………… 38
课后思考问题 …………………………………………………………………………… 38

第 5 章　ASP.NET 的 Web 应用程序 ………………………………………………… 40
5.1　ASP.NET 应用程序基础 ………………………………………………………… 40
　　5.1.1　网页间数据共享的基础 ……………………………………………………… 40
　　5.1.2　网页间的数据传递方法 ……………………………………………………… 41
5.2　Global.asax 文件的使用 ………………………………………………………… 41

		5.2.1 Global.asax 文件的结构	41
		5.2.2 Global.asax 文件的使用	43
	5.3	Application 对象的状态管理	44
		5.3.1 Application 对象基础	44
		5.3.2 网站的访客计数	46
	5.4	Session 对象的状态管理	47
		5.4.1 Session 对象的基础	47
		5.4.2 目前有多少人仍在线	48
	5.5	Cookie 的处理	49
		5.5.1 Cookie 基础	49
		5.5.2 如何使用 Cookie	50
	5.6	小结	53
	课后思考问题		53

第6章 ADO.NET 的应用 54

6.1	ADO.NET 对象模型	54
6.2	Connection 对象	55
6.3	Command 对象	57
	6.3.1 创建 Command 对象	57
	6.3.2 执行 Command 对象	57
	6.3.3 Command 对象实现添加功能	58
	6.3.4 参数查询	59
	6.3.5 Command 对象实现更新功能	60
	6.3.6 Command 对象实现删除功能	62
6.4	DataSet	63
	6.4.1 DataSet 的数据库操作	63
	6.4.2 ADO.NET 的 DataSet 数据模型	63
	6.4.3 DataSet 对象的三大特性	63
	6.4.4 DataSet 对象的数据库操作	64
	6.4.5 DataSet 对象实现插入操作	65
	6.4.6 更新记录	67
	6.4.7 删除记录	68
6.5	从数据表中获取单一字段值	69
6.6	DataReader 对象以表格显示数据表	70
6.7	DataSet 对象以表格显示数据表	72
6.8	小结	74
课后思考问题		74

第7章 数据控件 75

7.1	数据源控件	75

7.2 数据绑定 ……………………………………………………………… 78
　　7.2.1 数据绑定基础 ………………………………………………… 78
　　7.2.2 ListBox 控件的数据绑定 …………………………………… 79
7.3 数据列表控件 …………………………………………………………… 80
　　7.3.1 GridView 控件的常用事件 …………………………………… 80
　　7.3.2 使用 GridView 控件绑定数据源 ……………………………… 81
　　7.3.3 设置 GridView 控件的外观 …………………………………… 82
　　7.3.4 制定 GridView 控件的列 ……………………………………… 84
　　7.3.5 查看 GridView 控件中数据的详细信息 ……………………… 84
　　7.3.6 使用 GridView 控件分页显示数据 …………………………… 87
　　7.3.7 在 GridView 控件中实现全选和全不选功能 ………………… 87
　　7.3.8 在 GridView 控件中对数据进行编辑操作 …………………… 89
7.4 DataList 控件 ………………………………………………………… 90
　　7.4.1 DataList 控件概述 …………………………………………… 90
　　7.4.2 DataList 控件常用的属性、方法和事件 …………………… 91
　　7.4.3 使用 DataList 控件绑定数据源 ……………………………… 92
　　7.4.4 分页显示 DataList 控件中的数据 …………………………… 94
　　7.4.5 查看 DataList 控件中数据的详细信息 ……………………… 96
　　7.4.6 在 DataList 控件中对数据进行编辑操作 …………………… 99
　　7.4.7 获取 DataList 控件中控件数据的方法 ……………………… 101
　　7.4.8 在 DataList 控件中创建多个列 ……………………………… 102
7.5 Repeater 控件 ………………………………………………………… 102
　　7.5.1 Repeater 控件以表格显示数据表 …………………………… 103
　　7.5.2 Repeater 控件分页显示数据表中数据 ……………………… 104
7.6 小结 …………………………………………………………………… 107
课后思考问题 …………………………………………………………………… 107

第 8 章 单表新闻发布系统的实现 …………………………………………… 108
8.1 需求分析 ……………………………………………………………… 108
8.2 新闻标题显示 ………………………………………………………… 108
8.3 新闻具体内容的显示 ………………………………………………… 112
8.4 新闻检索功能 ………………………………………………………… 114
8.5 新闻后台登录页面实现 ……………………………………………… 115
8.6 添加新闻 ……………………………………………………………… 117
8.7 编辑新闻 ……………………………………………………………… 119
8.8 小结 …………………………………………………………………… 126

第 9 章 新闻发布系统 ………………………………………………………… 127
9.1 需求分析 ……………………………………………………………… 127
9.2 功能描述 ……………………………………………………………… 128

  9.2.1 后台登录 …………………………………………………………………… 128
  9.2.2 新闻栏目和类别管理 ……………………………………………………… 129
  9.2.3 用户管理 …………………………………………………………………… 129
  9.2.4 新闻发布 …………………………………………………………………… 129
  9.2.5 日志管理、流量统计及当日新闻查看 …………………………………… 129
  9.2.6 前台显示页面 ……………………………………………………………… 129
 9.3 数据库说明 ……………………………………………………………………… 130
 9.4 相应的主要代码 ………………………………………………………………… 131
  9.4.1 前台代码 …………………………………………………………………… 131
  9.4.2 后台代码 …………………………………………………………………… 148

参考文献 ……………………………………………………………………………… 162

第 1 章 ASP.NET 初识

本章要点
- ASP.NET 概述。
- ASP.NET 运行原理。
- ASP.NET 网站建设。

1.1 ASP.NET 简介

1.1.1 ASP.NET 发展历程

ASP.NET 是作为.NET 框架体系结构的一部分推出的。2000 年 ASP.NET 1.0 正式发布,2003 年 ASP.NET 升级为 1.1 版本。ASP.NET 1.1 发布之后更加激发了 Web 应用程序开发人员对 ASP.NET 的兴趣。于是在 2005 年 11 月微软公司又发布了 ASP.NET 2.0。ASP.NET 2.0 的发布是.NET 技术走向成熟的标志。ASP.NET 2.0 技术增加了大量方便、实用的新特性,是一种建立在公共语言运行库上的编程框架,可用于在服务器上开发功能强大的 Web 应用程序。它不但执行效率大幅度提高,对代码的控制也做得更好,并且支持 Web Controls 功能和多种语言,以高安全性、易管理性和高扩展性等特点著称。

ASP.NET 技术从 1.0 版本升级到 1.1 变化不是很大。从 ASP.NET 1.x 到 ASP.NET 2.0,却发生相当大的变化,在开发过程中微软公司深入市场,针对大量开发人员和软件使用者进行了卓有成效的研究,并为其指定了开发代号 ASP.NET Whidbey。ASP.NET 2.0 设计目标的核心可以用一个词"简化"来形容。因为其设计目标是将应用程序代码数减少 70%,改变过去那种需要编写很多重复性代码的状况,尽可能做到写很少的代码就能完成任务。对于应用构架师和开发人员而言,可以说 ASP.NET 2.0 是 Microsoft Web 开发史上的一个重要的里程碑。

1.1.2 ASP.NET 能做哪些事情

ASP.NET 作为 Web 设计技术的一种,它的主要功能是产生动态网页,以满足不同用户的需要。在网上见得最多的是 HTML 页面(文件的后缀名为.htm 或.html),它只是将别人编辑好的 Web 页面展现给大家,这种页面如果不手动去修改,是不会发生任何变化的。如果想在网页中访问数据库,这种纯 HTML 页面就无能为力了。这时就得依赖各种 Web 技术。聊天室、论坛等都是由这些技术产生的。

ASP.NET 作为一种新的 Web 技术,它给予设计者一种全新的 Web 设计概念。它将软件设计和 Web 设计融为一个整体。由于它和 VC.NET、VB.NET、VC# 这些程序设计语言使用同一个.NET Framework 对象开发库,可以想象 ASP.NET 所能实现的功能是多么强大,几乎 VC.NET 能做到的事情,ASP.NET 也能做到。

1.1.3 ASP.NET 的运行原理

当一个 HTTP 请求到达服务器并被 IIS 接收到之后,IIS 首先通过客户端请求的页面类型为其加载相应的 dll 文件,然后在处理过程中将这条请求发送给能够处理这个请求的模块。在 ASP.NET 2.0 中,这个模块称为 HttpHandler(HTTP 处理程序组件),之所以 .aspx 这样的文件可以被服务器处理,就是因为在服务器端有默认的 HttpHandler 专门处理 .aspx 文件。IIS 再将这条请求发送给能够处理这个请求的模块之前,还需要经过一些 HttpModule 的处理,这些都是系统默认的 Modules(用于获取当前应用程序的模块集合),在这个 HTTP 请求转到 HttpHandler 之前要经过不同的 HttpModuls 的处理。这样做的好处,一是为了一些必需的过程,二是为了安全性,三是为了提高效率,四是为了用户能够在更多的环节上进行控制,增强用户的控制能力。ASP.NET 2.0 的运行原理如图 1.1 所示。

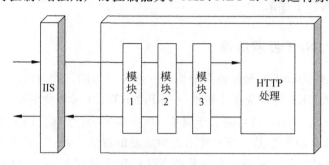

图 1.1　ASP.NET 2.0 运行原理

① 用户请求发送到 Web 服务器；② 将请求通过相应的.dll 文件发送到 ASP.NET 引擎；
③ 用户请求通过 HTTP 模块；④ HTTP 处理模块被调用,并返回到用户请求文件

1.2 制作一个 ASP.NET 网站

1.2.1 创建 ASP.NET 网站

启动 Visual Studio 2005 开发环境后,首先进入"起始页"界面,执行"文件"→"新建"→"网站"命令,创建一个 ASP.NET 网站,如图 1.2 所示。

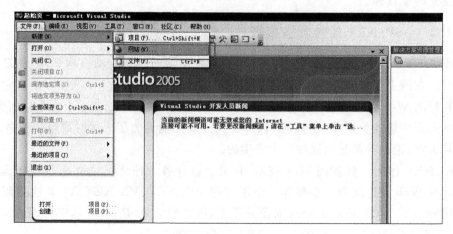

图 1.2　新建网站

选择创建"网站"后,在打开的"新建网站"窗口中,选择"ASP.NET 网站",并输入网站的位置,单击"确定"按钮,完成 test 网站的创建,如图 1.3 所示。

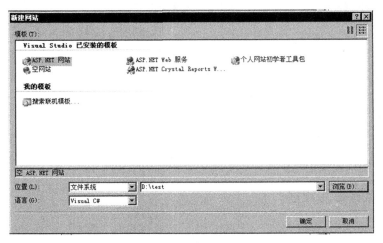

图 1.3 新建网站窗口

此时会进入 Visual Studio 2005 Web 页设计页面,Web 窗体的布局如图 1.4 所示。

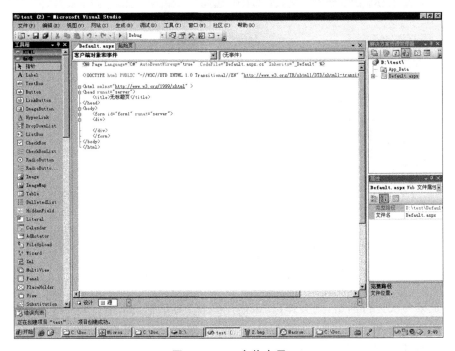

图 1.4 Web 窗体布局

1.2.2 写入代码

创建了一个网站后,接下来的工作就是设计 Web 页面。一个 Web 页包含两部分:"设计"视图和"源"视图。在"设计"视图中,用户可以从"工具箱"选项卡中直接选择各种控件添加到 Web 页面上,也可在页面上直接输入文字。

在"源"视图中,直接写入代码,如图 1.5 所示。

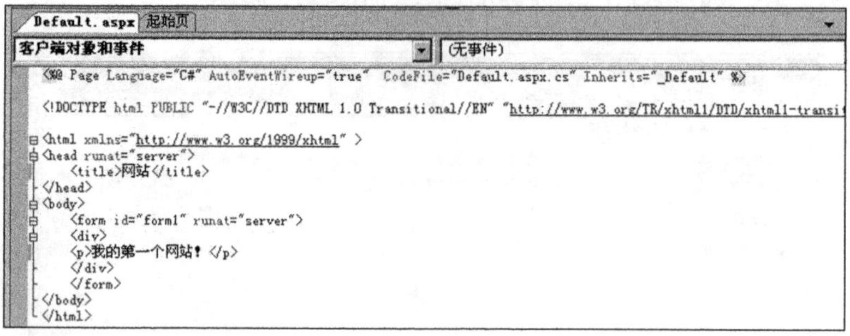

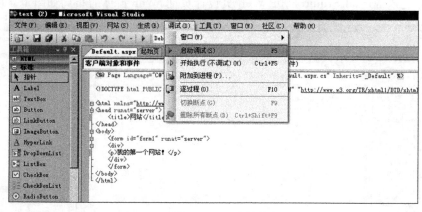

图 1.5　在"源"视图中加入代码

1.2.3　运行应用程序

Visual Studio 中有多种方法运行应用程序。可以执行"调试"→"启动调试"命令运行应用程序,如图 1.6 所示;也可以单击工具栏上的"调试"按钮运行应用程序,如图 1.7 所示;还可以直接按 F5 键运行程序。

图 1.6　启动"调试"菜单运行程序

图 1.7　通过工具栏运行应用程序

如果是第一次运行网站,则会弹出"未启用调试"对话框,如图 1.8 所示。该对话框中,用户可以选择添加 Web.config 文件。单击"确定"按钮,就可以将 Web.config 文件添加到 Web 应用程序中,并启用调试功能。

最后运行的结果如图 1.9 所示。

这是个简单的例子,细心的读者会发现,在这个例子中除了后缀名为.aspx 的文件以外,根本没有用到一个属于 ASP.NET 的特性。实际上这个例子的程序代码就只是纯 HTML,没有一点延伸。这个例子反映了一个特性,那就是任何一个纯 HTML 文件都可以改写成为 ASP.NET 程序。

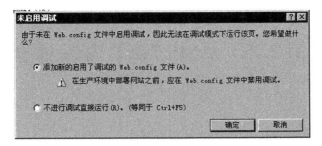

图 1.8 "未启用调试"对话框

图 1.9 运行结果

1.3 小结

本章主要介绍 ASP.NET 的运行原理,以及如何在 Visual Studio 中建立网站。通过学习,读者能熟悉 Visual Studio 的主要界面,了解 Visual Studio 的运行机制。

课后思考问题

1. 如何在 Web 应用中再添加一个页面?
2. web.config 文件能否在 Web 应用程序中新建?如何建立?
3. web.config 文件的作用是什么?

第 2 章 ASP.NET 内置对象

本章要点
- 了解 Response、Request、Server 对象的用法。
- 初步了解 ASP.NET 的输入输出方法。

本章将通过介绍 ASP.NET 中的 Response、Request、Server 这 3 个常用对象的应用，学习 ASP.NET 处理动态 Web 页面的一般方法。

2.1 Response 对象

Response 对象是实际在执行 System.Web 命名空间中的类 HttpResponse。CLR 会根据用户的请求信息建立一个 Response 对象，Response 将用于回应客户浏览器，告诉浏览器回应内容的报头、服务器端的状态信息以及输出指定的内容。

1. Response.Write() 方法

Write 方法输出指定的文本内容。例如：

```
Response.Write("欢迎来到ASP.COM");
```

Write 方法还可以输出 HTML 代码，以字符串的形式输出。例如：

```
Response.Write("<a href='http://www.sina.com'>新浪</a>");
```

2. Response.End() 方法

End 方法使得 Web 服务器停止当前程序的处理并返回结果。剩下文件内容是没有处理的。这个方法可以用在调试过程中使用。

3. Response.Redirect(url) 方法

Redirect 方法会让页面自动的跳转到 url 所指定的页面。在 Default.aspx.cs 页面中输入如下代码：

```
protected void Page_Load(object sender, EventArgs e)
{
    Response.Redirect("~\\TansferTest.aspx");
}
```

运行 Default.aspx 页面，最终的运行结果如图 2.1 所示。

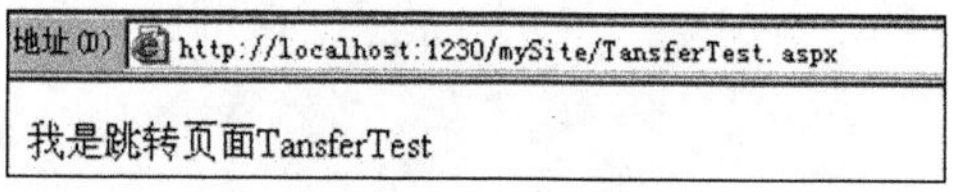

图 2.1 运行结果

2.2 Request 对象

Request 对象实际上操作 System.Web 命名空间中的类 HttpRequest。当客户发出请求执行 ASP.NET 程序时,CLR 会将客户端的请求信息包装在 Request 对象中。这些请求信息包括请求报头(Header)、请求方法(如 POST、GET)、参数名和参数值等。

Request 对象的调用方法如下:

```
Request.Collection["Variable"]
```

其中,Collection 包括 QueryString、Form、Cookies、ServerVariables 4 种集合。这里的 Collection 可以省略,也就是说 Request["Variable"]与 Request.Collection["Variable"]这两种写法都是允许的。如果省略了 Collection,那么 Request 对象会依照 QueryString、Form、Cookies、ServerVariables 的顺序查找,直至发现 Variable 所指的关键字并返回其值,如果没有发现其值,方法则返回空值(Null)。

为了优化程序的执行效率,建议最好还是使用 Collection,如果过多地搜索,会降低程序的执行效率。

2.2.1 QueryString 集合

QueryString 集合收集的信息来自于请求 URL 地址中"?"号后面的数据,这些数据称为"URL 附加信息":

```
http://www.aspcn.com/show.asp?id=111
```

在此 URL 中,QueryString 收集到的附加信息是"show.asp?"后的数据 id=111。

此时,取得参数 id 的参数值的语句如下:

```
Request.QueryString["id"]
```

QueryString 主要用于收集 HTTP 中的 GET 请求发送的数据,如果在一个请求事件中被请求的程序 URL 地址出现了"?"号的数据,则表示此次的请求方式为 GET。GET 方法是 HTTP 中的默认请求方法,最常用的超链接,便是 GET 方法发送请求。

直接在 URL 中写上参数与参数值发送 GET 请求:

```
<a href="show.aspx?id=111">显示 ID 为 111 的文章</a>
```

这样就可以在 show.aspx 页面通过 Request.QueryString["id"]方式获取到地址栏中传过来的 id 的值。

也可以通过 Form 来发送 GET 请求。

```
<form action="show.aspx" method="get">
    <input type="text" name="id" value="111">
</form>
```

<form>标签中的 method 并没有设置,因为浏览器默认的发送方法就是 GET,并不需要明确标明,当然标明了也不会出错。

下面通过两个实例来演示使用 Form 发送 GET 请求。演示将使用两个名为 login.

html 和 login.aspx 的文件。

login.html 如下：

```
<html>
<head>
    <title>登录页面</title>
</head>
<body>
    <form action="login.aspx" post="get">
    用户名：<input id="username" name="username" type="text" /><br />
    密 码：<input id="pwd" name="pwd" type="password" /><br />
    <input id="Submit1" type="submit" value="提交" />
    <input id="Reset1" type="reset" value="重置" />
    </form>
</body>
</html>
```

login.aspx 如下：

```
<html>
<head runat="server">
    <title>接受 GET 请求</title>
</head>
<body>
    <h2>接受 GET 请求</h2>
    <%
    string username =Request.QueryString["username"];
    string pwd =Request.QueryString["pwd"];
    Response.Write("你的用户名为：" +username +"<br/>密码为：" +pwd);
    %>
</body>
</html>
```

运行结果如图 2.2 和图 2.3 所示。

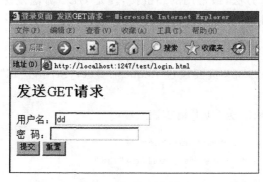

图 2.2　发送 GET 请求

GET 方法会将传递的参数与参数值添加到 URL 地址中，而且包含附加信息的 URL 地址均会显示在浏览器的地址栏中。

图 2.3 接受 GET 请求

2.2.2 Form 集合

GET 方法是将传递的数据追加至 URL 中。URL 地址长度是有限制的,因此使用 GET 方法所能传递的数据也是有限。一般地,GET 方法能够传递 256B 的数据。在多数情况下,GET 方法传递的数据长度是远远不够的。这时便需要使用 HTTP 的另外一种请求方式 POST,POST 方法可传递的数据最大值为 2MB。

POST 请求必须由 Form 发出,例如:

```
<form action="login.aspx" method="post">
    您的大名:<input type="text" name="username" /><br/>
    <input type="submit" value="发送" />
<form>
```

使用 POST 请求时,<form>标签中的 method 属性值设置为 POST。
ASP.NET 使用 Request.Form 方法接收 POST 方法传递的数据。

```
Request.Form["variable"]
```

下面通过实例来演示 POST 提交,演示将使用两个名为 login.html 和 login.aspx 的文件。
login.html 如下:

```
<html>
<head>
    <title>登录页面 发送 POST 请求</title>
</head>
<body>
    <h2>发送 POST 请求</h2>
    <form action="login.aspx" method="post">
        用户名:<input id="username" name="username" type="text" /><br />
        密 码:<input id="pwd" name="pwd" type="password" /><br />
        <input id="Submit1" type="submit" value="提交" />
        <input id="Reset1" type="reset" value="重置" />
    </form>
</body>
</html>
```

login.aspx 如下:

```
<html>
<head runat="server">
    <title>接受 POST 请求</title>
</head>
<body>
    <h2>接受 POST 请求</h2>
    <%
    string username =Request.Form["username"];
    string pwd =Request.Form["pwd"];
    Response.Write("你的用户名为: " +username +"<br/>密码为: " +pwd);
    %>
</body>
</html>
```

运行结果如图 2.4 和图 2.5 所示。

图 2.4　发送 POST 请求

图 2.5　接受 POST 请求

从图 2.4 可以看出，通过 POST 方式发送的数据将不会显示在 URL 中，因此用 POST 发送数据会比 GET 发送数据安全。

2.3　Server 对象

　　Response 对象是执行 System.Web 命名空间中的类 HttpServerUtility。Server 对象提供许多访问的方法和属性帮助程序有序地执行。

1. Server.MapPath() 方法

Server.MapPath 方法将虚拟路径转换为绝对路径。例如在.cs 页面中输入以下代码：

```
Response.Write(Server.MapPath("~\\"));
```

会在屏幕上输出当前网站的绝对路径 D:\mySite\。此绝对路径就是当前网站的根目录，就是解决方案中显示的根目录，如图 2.6 所示。

2. Server.Transfer(path) 方法

终止当前程序的执行，自动跳转到 path 所指的页面。在 Default.aspx.cs 页面中输入如下代码：

```
protected void Page_Load(object sender, EventArgs e)
{
    Server.Transfer("~\\TansferTest.aspx");
}
```

运行后结果如图 2.7 所示。

图 2.6　解决方案中显示的网站路径

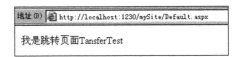

图 2.7　运行结果

如图 2.7 所示，地址栏中显示的是 Default.aspx，但是页面上显示的是 TansferTest.aspx 页面的内容，页面已经自动的跳转到 TansferTest.aspx 页中。这也能看出 Server.Transfer(path) 与 Response.Redirect(url) 两个方法的不同之处。

2.4　小结

本章主要讲述 Request 对象、Response 对象和 Sever 对象的使用方式以及 3 个对象对应的属性和方法等。

课后思考问题

1. 什么是相对路径？什么是绝对路径？在网站的开发中多用什么路径，是相对的路径还是绝对的路径。
2. Server.MapPath("~\\")中 ~ 代表什么意思？
3. "Server.Transfer("~\\TansferTest.aspx");"这个路径写法为什么得用 ~\\？不用会怎么样？

第3章 标准控件

本章要点
- 使用 Label 控件。
- 使用 TextBox 控件。
- 使用 DropDownList 控件。
- 使用 ListBox 控件。
- 使用 CheckBox 控件。
- 使用 RadioButton 控件。
- 使用 Image 控件。
- 使用 Button 控件、LinkButton 控件和 ImageButton 控件。
- 使用 HyerLink 控件。
- 使用 FileUpload 控件上传文件。
- 使用 Calendar 控件。

3.1 Label 控件

3.1.1 Label 控件的属性

Label 控件又称为标签控件，主要用来显示文本信息，此控件属于服务器端标准控件。label 控件的主要属性有 ID 和 Text。

(1) ID：表示 Label 控件的 ID 名称。
(2) Text：表示 Label 控件显示的文本。

3.1.2 Label 控件的实际使用

本实例使用 Label 控件显示文本。
主要实现步骤如下。

(1) 新建一个网站，在默认主页 Default.aspx 页面中添加一个 Label 控件。如图 3.1 所示，双击工具箱中的 Label 控件，就会在 Default.aspx 页面的设计视图添加一个 Label 控件，如图 3.2 所示。

(2) 在设计视图中，选中 Label 控件，在属性工具栏中设置 Label 控件的相应属性，如图 3.3 和图 3.4 所示。

(3) 也可以在代码视图中设置 Label 控件的属性，代码如下：

```
protected void Page_Load(object sender, EventArgs e)
{
    Label1.Text="显示 Label 文本";
}
```

图 3.1 在工具箱中使用 Label 控件

图 3.2 在页面中添加 Label 控件

图 3.3 设置 Label 控件的 ID 属性

图 3.4 设置 Label 控件的 Text 属性

（4）最终运行的结果如图 3.5 所示。

图 3.5 Label 控件显示的文本

3.2 TextBox 控件

3.2.1 TextBox 控件基本属性

在 Web 页面中，常常使用文本框控件（TextBox）来接受用户的输入信息，包括文本、数字和日期等。默认情况下，文本框控件是一个单行的文本框，用户只能输入一行内容。但是通过修改它的属性，可以将文本框改为允许输入多行文本或者输入密码的形式，此控件属于服务器端标准控件。它的主要属性包括 ID、Text 和 TextMode。

（1）ID：表示 TextBox 控件的 ID 名称。

（2）Text：表示 TextBox 控件显示的文本。

（3）TextMode：用于控制 TextBox 控件的文本显示方式，该属性的设置选项有以下 3 种。

① 单行(SingleLine)：用户只能在一行中输入信息，还可以选择限制控件接收的字符数。

② 多行(MultiLine)：文本很长时，允许用户输入多行文本并执行换行。

③ 密码(Password)：将用户输入的字符用黑点屏蔽，以隐藏这些信息。

3.2.2　TextBox 控件的实际使用

本实例通过设置 TextBox 控件的 TextMode 属性来实现该控件的 3 种文本显示效果，并通过代码把页面上填写的内容输出出来。

实现步骤如下。

（1）新建一个网站，在默认主页 Default.aspx 页面中添加 3 个 TextBox 控件，如图 3.6 所示。

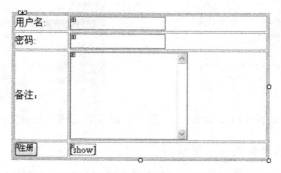

图 3.6　页面添加的 3 个 TextBox 控件

（2）在属性窗口中分别设置 TextBox 的属性如下。

① 输入用户名的 TextBox 控件的 TextMode 属性为 SingleLine。

② 输入密码的 TextBox 控件的 TextMode 属性为 Password。

③ 输入备注的 TextBox 控件的 TextMode 属性为 MultiLine。

设计页面 TextBoxTest.aspx 的代码如下：

```
<div>
    <table style="width: 419px">
    <tr>
      <td>用户名:</td>
      <td><asp:TextBox ID="username" runat="server"></asp:TextBox></td>
    </tr>
    <tr>
      <td>密码:</td>
      <td><asp:TextBox ID="pwd" runat="server" TextMode="Password"></asp:
          TextBox></td>
    </tr>
```

```
        <tr>
          <td>备注:</td>
          <td><asp:TextBox ID="remark" runat="server" Height="136px" TextMode=
              "MultiLine" Width="186px"></asp:TextBox></td>
        </tr>
        <tr>
          <td><asp:Button ID="btn" runat="server" Text="注册" OnClick="btn_Click" /></
          td>
          <td><asp:Label ID="show" runat="server" Text=""></asp:Label></td>
        </tr>
      </table>
</div>
```

代码区域 TextBoxTest.aspx.cs 的代码如下:

```
protected void btn_Click(object sender, EventArgs e)
{
    show.Text ="用户名为: " +username.Text +"<br/>密码为: " +pwd.Text +"<br/>备注
        信息为: " +remark.Text;
}
```

最终运行的结果如图 3.7 所示。

图 3.7　运行结果

3.3　DropDownList 控件

3.3.1　DropDownList 控件添加列表项

　　DropDownList 控件是人们经常使用的下拉列表,用于收集用户的选择信息。通常一个 DropDownList 控件创建一个包含多个选项的下拉列表,用户可以从中选择一个选项。

使用 DropDownList 控件时,有两种方法可以指定 DropDownList 控件的列表项:一种是通过逐个输入来输入列表项;另一种则是通过绑定数据库的方法指定列表项,这种方法将在第 7 章介绍。

要向 DropDownList 控件中添加列表项,就要编辑它的 Item 属性。选中当前的 DropDownList 控件,在如图 3.8 所示的属性窗口中,滚动到 Items 属性。可以看到该属性值当前为 Collection。如果单击该属性值,一个矩形的按钮将出现在 Collection 右边。

单击该按钮,打开"ListItem 集合编辑器"对话框,如图 3.9 所示。通过该对话框可以向 DropDownList 控件添加、编辑或删除列表项。

图 3.8　向 DropDownList 控件中添加列表项　　图 3.9　"ListItem 集合编辑器"对话框

单击"添加"按钮,在"ListItem 集合编辑器"对话框左边的文本框中将添加一个新条目,在右边将显示新添加列表项的属性。如图 3.10 所示,列表项有 4 个属性:Enabled、Selected、Text 和 Value。

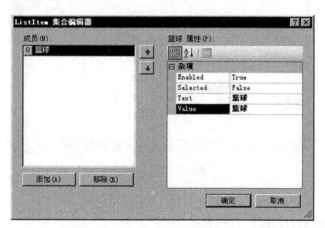

图 3.10　"ListItem 集合编辑器"对话框

(1) Enabled 属性指定列表项是否能最终在下拉列表中出现。

(2) Selected 属性指定在网页被加载的时候该列表项是否默认被选择。

(3) Text 属性是指下拉列表中显示给用户的文本。

(4) Value 属性对用户来说是不可见的,它只是用作一种传送列表项相关信息的方法,并不希望用户看到这些信息。

注意:如果不指定 DropDownList 控件列表项的 Value 属性,系统默认将 Value 属性和 Text 属性设为相同的值。

3.3.2 获取 DropDownList 控件列表项的值

前面介绍的 DropDownList 控件的列表项有 4 个属性:Enabled、Selected、Text 和 Value。因此,要获取 Text 属性,可以使用以下方法:

```
DropDownListID.SelectedItem.Text
```

同样,要获取 Value 属性,可以使用下述方法:

```
DropDownListID.SelectedItem.Value
```

3.3.3 DropDownList 控件的实际应用

本实例通过设置 DropDownList 控件的 Item 属性来实现该控件的显示效果,并通过代码把页面上选择的内容输出出来。

实现步骤如下。

(1) 新建一个网站,在默认主页 Default.aspx 页面中添加 1 个 DropDownList 控件。

(2) 在属性窗口中分别设置 DropDownList 的属性,如图 3.11 所示。

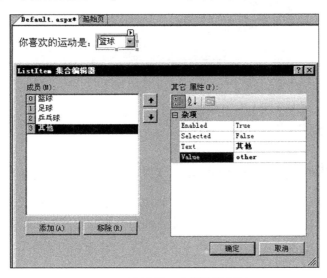

图 3.11 设计视图中设计的页面

设计页面 Default.aspx 的代码如下:

```
<div>
    你喜欢的运动是:<asp:DropDownList ID="hobby" runat="server">
        <asp:ListItem Value="basketball">篮球</asp:ListItem>
        <asp:ListItem Value="football">足球</asp:ListItem>
```

```
        <asp:ListItem Value="pingpang">乒乓球</asp:ListItem>
        <asp:ListItem Value="other">其他</asp:ListItem>
    </asp:DropDownList>
    <br />
    <asp:Button ID="Button1" runat="server" Text="请选择" OnClick="Button1_Click" />
    <br />
    <asp:Label ID="show" runat="server"></asp:Label>
</div>
```

代码区域 Default.aspx.cs 代码如下：

```
protected void Button1_Click(object sender, EventArgs e)
{
    show.Text = "你喜欢的运动是: " +hobby.SelectedItem.Text;
    show.Text += "<br/>你选择的运动的value值为: " +hobby.SelectedItem.Value;
}
```

最终运行的结果如图 3.12 所示。

图 3.12　使用下拉列表框的运行结果

3.4　ListBox 控件

3.4.1　ListBox 控件添加列表项

ListBox 控件就是人们常用的列表框，它的用法以及属性和 DropDownList 控件非常相似。本节用 ListBox 控件实现和 3.3.3 节利用 DropDownList 控件的示例差不多的功能。

3.4.2　ListBox 控件实际使用

本实例通过设置 ListBox 控件的 Item 属性来实现该控件的显示效果，并通过代码把页面上选择的内容输出出来。

实现步骤如下。

（1）新建一个网站，在默认主页 ListBox.aspx 页面中添加 1 个 ListBox 控件。

（2）在属性窗口中分别设置 ListBox 控件的 Items 属性。

设计页面 ListBox.aspx 的代码如下：

```
<div>
    你喜欢的运动是：<asp:ListBox ID="play" runat="server">
        <asp:ListItem>篮球</asp:ListItem>
        <asp:ListItem>足球</asp:ListItem>
        <asp:ListItem>乒乓球</asp:ListItem>
        <asp:ListItem>其他</asp:ListItem>
        <asp:ListItem></asp:ListItem>
    </asp:ListBox>
    <asp:Button ID="submit" runat="server" Text="确定" onclick="submit_Click" />
    <br />
    <asp:Label ID="show" runat="server"></asp:Label>
</div>
```

代码区域 ListBox.aspx.cs 代码如下：

```
protected void submit_Click(object sender, EventArgs e)
{
    show.Text ="你喜欢的运动是：" +play.SelectedItem.Text +"<br/>" +"你喜欢的运动是："+play.SelectedItem.Value;
}
```

最终运行的结果如图 3.13 所示。

图 3.13 使用 ListBox 控件的运行结果

3.4.3 ListBox 控件实现多项选择

ListBox 控件比 DropDownList 控件多了可以多选的功能，设置 ListBox 控件的 SelectionMode 属性为 Multiple，就能实现多选了。在选择的时候，按住键盘上的 Ctrl 键再选择，以实现多项选择。

本实例通过设置 ListBox 控件的 SelectionMode 属性来实现该控件的多项选择功能，并通过代码把页面上选择的内容输出出来。

设计页面 ListBox.aspx 的代码如下：

```
<div>
    你喜欢的运动是：<asp:ListBox ID="play" runat="server" SelectionMode="Multiple">
```

```
        <asp:ListItem>篮球</asp:ListItem>
        <asp:ListItem>足球</asp:ListItem>
        <asp:ListItem>乒乓球</asp:ListItem>
        <asp:ListItem>其他</asp:ListItem>
        <asp:ListItem></asp:ListItem>
    </asp:ListBox>
    <asp:Button ID="submit" runat="server" Text="确定" onclick="submit_Click" />
    <br />
    <asp:Label ID="show" runat="server"></asp:Label>
</div>
```

代码区域 ListBox.aspx.cs 代码如下：

```
protected void submit_Click(object sender, EventArgs e)
{
    show.Text ="你喜欢的运动是：";
    for (int i =0; i <play.Items.Count; i++)
    { if (play.Items[i].Selected)
      { show.Text +=play.Items[i].Text +"<br/>";
      }
    }
}
```

最终运行的结果如图 3.14 所示。

图 3.14　使用可以多选的 ListBox 控件的运行结果

3.5　CheckBox 控件

3.5.1　CheckBox 控件基本属性

CheckBox 控件用于在 ASP.NET 中添加一个复选框。主要属性有 ID 和 Checked，可以使用 CheckBoxID.Checked 属性来指定，它将返回一个 True 和 False 值，来确定该复选框有没有被选定。

下面用一个选择运动的例子来说明 CheckBox 控件的使用。

设计页面 CheckBox.aspx 的代码如下：

```
<div>
    你喜欢的运动是：<br />
    <asp:CheckBox ID="bask" runat="server" Text="篮球" /><br />
    <asp:CheckBox ID="foot" runat="server" Text="足球" /><br />
    <asp:CheckBox ID="ping" runat="server" Text="乒乓球" /><br />
    <asp:CheckBox ID="other" runat="server" Text="其他" /><br />
    <asp:Button ID="submit" runat="server" OnClick="submit_Click" Text="确定" />
    <br />
    <asp:Label ID="show" runat="server"></asp:Label>
</div>
```

代码区域 CheckBox.aspx.cs 代码如下：

```
protected void submit_Click(object sender, EventArgs e)
{
    show.Text ="你喜欢的运动是：";
    if (bask.Checked)
    {   show.Text +=bask.Text +"<br/>";
    }
    if (foot.Checked)
    {   show.Text +=foot.Text +"<br/>";
    }
    if (ping.Checked)
    {   show.Text +=ping.Text +"<br/>";
    }
    if (other.Checked)
    {   show.Text +=other.Text +"<br/>";
    }
}
```

最终的运行结果如图 3.15 所示。

图 3.15 使用 CheckBox 控件的运行结果

3.5.2 CheckBoxList 控件

为了方便复选控件的使用,.NET 服务器控件中包括了复选组控件,拖动一个复选组控件到页面可以添加复选组列表。

在网页 CheckBoxList.aspx 中加入 CheckBoxList 控件,编辑 CheckBoxList 控件的 Items 属性,添加相应的选项。添加步骤同 DropDownList 控件。

设计页面 CheckBoxList.aspx 的代码如下:

```
<div>
    你喜欢的运动是:<br />
    <asp:CheckBoxList ID="play" runat="server">
        <asp:ListItem>篮球</asp:ListItem>
        <asp:ListItem>足球</asp:ListItem>
        <asp:ListItem>乒乓球</asp:ListItem>
        <asp:ListItem>其他</asp:ListItem>
    </asp:CheckBoxList><br />
    <asp:Button ID="submit" runat="server" OnClick="submit_Click" Text="确定" /><br />
    <asp:Label ID="show" runat="server"></asp:Label>
</div>
```

代码区域 CheckBoxList.aspx.cs 代码如下:

```
protected void submit_Click(object sender, EventArgs e)
{
    show.Text ="你喜欢的运动是:";
    for (int i =0; i <play.Items.Count; i++)
    {   if (play.Items[i].Selected)
        {   show.Text +=play.Items[i].Text +"<br/>";
        }
    }
}
```

最终的运行结果如图 3.16 所示。

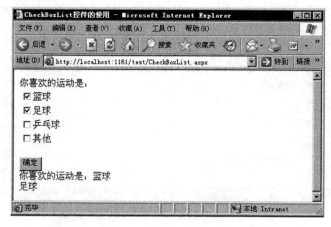

图 3.16 使用 CheckBoxList 控件的运行结果

3.6 RadioButton 控件

3.6.1 RadioButton 控件基本属性

RadioButton 控件用于在 ASP.NET 网页中添加单选按钮,用于从一个有效选项列表中选择一个选项。可以将单选按钮编成相关的组。给定一组相关按钮,不能同时选中多个单选按钮,也就是说,相关按钮互斥。在网页中加入 RadioButton 控件时,加入的多个 RadioButton 控件并不互斥,也就是说可以选择多个单选按钮,如何实现一次只能选择一个单选按钮,这就需要设置 RadioButton 控件的 GroupName 属性。单选控件常用属性如下所示。

(1) Checked:控件是否被选中。
(2) GroupName:单选控件所处的组名。

单选控件通常需要 Checked 属性来判断某个选项是否被选中,多个单选控件之间可能存在着某些联系,这些联系通过 GroupName 进行约束和联系。

本实例通过设置 RadioButton 控件的相应属性来实现该控件的选择功能,并通过代码把页面上选择的内容输出出来。

设计页面 RadioButton.aspx 的代码如下:

```
<div>
    你喜欢的运动是:<br />
    <asp:RadioButton ID="bask" runat="server" GroupName="sport" Text="篮球" />
    <asp:RadioButton ID="foot" runat="server" GroupName="sport" Text="足球" />
    <asp:RadioButton ID="ping" runat="server" GroupName="sport" Text="乒乓球" />
    <asp:RadioButton ID="other" runat="server" GroupName="sport" Text="其他" />
    <br />
    <asp:Button ID="submit" runat="server" onclick="submit_Click" Text="确定" />
    <br />
    <asp:Label ID="show" runat="server"></asp:Label>
</div>
```

代码区域 RadioButton.aspx.cs 代码如下:

```
protected void submit_Click(object sender, EventArgs e)
{
    if (bask.Checked)
    {   show.Text ="你喜欢的运动是篮球。";
    }
    else
        if (foot.Checked)
        {
            show.Text ="你喜欢的运动是足球。";
        }
```

```
            else
                if (ping.Checked)
                {   show.Text ="你喜欢的运动是乒乓球。";
                }
                else
                    if (other.Checked)
                    {   show.Text ="你喜欢的运动是其他。";
                    }
                    else
                    {   show.Text ="你没选择任何项。";
                    }
}
```

最终的运行结果如图 3.17 所示。

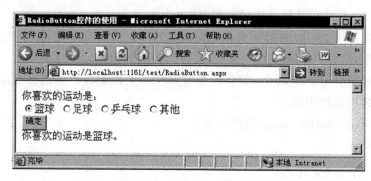

图 3.17 使用 RadioButton 控件的运行结果

3.6.2 RadioButtonList 控件

与单选控件相同,单选组控件也是只能选择一个项目的控件,而与单选控件不同的是,单选组控件没有 GroupName 属性,但是却能够列出多个单选项目。另外,单选组控一件所生成的代码也比单选控件实现的相对较少。

在网页 RadioButtonList.aspx 中加入 RadioButtonList 控件,编辑 RadioButtonList 控件的 Items 属性,添加相应的选项。添加步骤同 DropDownList 控件。

设计页面 RadioButtonList.aspx 的代码如下:

```
<div>
    你喜欢的运动是:<br />
    <asp:RadioButtonList ID="sport1" runat="server">
        <asp:ListItem>篮球</asp:ListItem>
        <asp:ListItem>足球</asp:ListItem>
        <asp:ListItem>乒乓球</asp:ListItem>
        <asp:ListItem>其他</asp:ListItem>
    </asp:RadioButtonList>
    <asp:Button ID="submit1" runat="server" onclick="submit1_Click" Text="确
        定" />
```

```
        <asp:Label ID="show1" runat="server"></asp:Label>
        <br />
</div>
```

代码区域 RadioButtonList.aspx.cs 代码如下：

```
protected void submit1_Click(object sender, EventArgs e)
{
    show1.Text = "你喜欢的运动是" + sport1.Text;
}
```

最终的运行结果如图 3.18 所示。

图 3.18 使用 RadioButtonList 控件的运行结果

3.7 Image 控件

图像控件用来在 Web 窗体中显示图像，图像控件常用的属性如下。

(1) AlternateText：在图像无法显式时显示的备用文本。
(2) ImageAlign：图像的对齐方式。
(3) ImageUrl：要显示图像的 URL。

当图片无法显示的时候，图片将被替换成 AlternateText 属性中的文字，ImageAlign 属性用来控制图片的对齐方式，而 ImageUrl 属性用来设置图像链接地址。同样，HTML 中也可以使用<img src="" alt="">来替代图像控件，图像控件具有可控性的优点，就是通过编程来控制图像控件，图像控件基本声明代码如下：

```
<asp:Image ID="Image1" runat="server" />
```

除了显示图形以外，Image 控件还有一些其他属性（可以为图像指定各种文本）：

(4) ToolTip：浏览器显式在工具提示中的文本。
(5) GenerateEmptyAlternateText：如果将此属性设置为 True，则呈现的图片的 alt 属性将设置为空。

开发人员能够为 Image 控件配置相应的属性以便在浏览时呈现不同的样式,创建一个 Image 控件也可以直接通过编写 HTML 代码进行呈现,示例代码如下:

```
<asp:Image ID="Image1" runat="server" AlternateText="图片连接失效" ImageUrl=
"~/images/cms.jpg" />
```

上述代码设置了一个图片,并当图片失效的时候提示图片连接失效。

注意:当双击图像控件时,系统并没有生成事件所需要的代码段,这说明 Image 控件不支持任何事件。

3.8 按钮控件

3.8.1 按钮控件的属性

在 Web 应用程序和用户交互时,常常需要提交表单、获取表单信息等操作。在这期间,按钮控件是非常必要的。按钮控件能够触发事件,或者将网页中的信息回传给服务器。在 ASP.NET 中,包含三类按钮控件,分别为 Button、LinkButton 和 ImageButton。

下面的语句声明了 3 种按钮,示例代码如下:

```
<asp:Button ID="Button1" runat="server" Text="Button" />    //普通的按钮<br />
<asp:LinkButton ID="LinkButton1" runat="server">LinkButton</asp:LinkButton>
                                                            //Link 类型的按钮<br />
<asp:ImageButton ID="ImageButton1" runat="server" />        //图像类型的按钮
```

对于 3 种按钮,它们起到的作用基本相同,主要是表现形式不同,如图 3.19 所示。

3.8.2 按钮控件的 Click 事件

这 3 种按钮控件对应的事件通常是 Click 单击事件。在 Click 单击事件中,通常用于编写用户单击按钮时所需要执行的事件,示例代码如下:

```
protected void Button1_Click(object sender,
EventArgs e)
{ Label1.Text ="普通按钮被触发";        //输出信息
}
protected void LinkButton1 _ Click ( object
sender, EventArgs e)
{ Label1.Text ="链接按钮被触发";        //输出信息
}
protected void ImageButton1_Click(object sender, ImageClickEventArgs e)
{ Label1.Text ="图片按钮被触发";        //输出信息
}
```

图 3.19 3 种按钮类型

上述代码分别为 3 种按钮生成了事件,其代码都是将 Label1 的文本设置为相应的文本,运行结果如图 3.20 所示。

图 3.20 按钮的 Click 事件

3.9 HyperLink 控件

超链接控件相当于实现了 HTML 代码中的<a href=""></a>效果,当然,超链接控件有自己的特点,当拖动一个超链接控件到页面时,系统会自动生成控件声明代码,示例代码如下:

```
<asp:HyperLink ID="HyperLink1" runat="server">HyperLink</asp:HyperLink>
```

上述代码声明了一个超链接控件,相对于 HTML 代码形式,超链接控件可以通过传递指定的参数来访问不同的页面。当触发了一个事件后,超链接的属性可以被改变。超链接控件通常使用的两个属性如下。

(1) ImageUrl:要显式图像的 URL。
(2) NavigateUrl:要跳转的 URL。

1. ImageUrl 属性

设置 ImageUrl 属性可以设置这个超链接是以文本形式显式还是以图片文件显式,示例代码如下:

```
<asp:HyperLink ID="HyperLink1" runat="server"
    ImageUrl="~/images/cms.jpg">
    HyperLink
</asp:HyperLink>
```

上述代码将文本形式显示的超链接变为了图片形式的超链接,虽然表现形式不同,但是不管是图片形式还是文本形式,全都实现的相同的效果。

2. Navigate 属性

Navigate 属性可以为无论是文本形式还是图片形式的超链接设置超链接属性,即即将跳转的页面,示例代码如下:

```
<asp:HyperLink ID="HyperLink1" runat="server"
```

```
            ImageUrl="~/images/cms.jpg"
            NavigateUrl="http://www.sina.com">
            HyperLink
        </asp:HyperLink>
```

上述代码使用了图片超链接的形式。其中图片来自本地网站/images/cms.jpg,当单击此超链接控件后,浏览器将跳到 URL 为 http://www.sina.com 的页面。

3.10 FileUpload 控件

在开发网站的时候,有时候要求网站或者系统有上传文件的功能,这样可以直接将文件上传到服务器中。在 ASP.NET 中,通过使用 FileUpload 控件可以方便地实现文件的上传。

首先要了解 FileUpload 控件的几个相关的属性和方法。

(1) 通过 FileUploadID.PostedFile.FileName 可以获得通过 FileUpload 控件选择的文件的完整的文件名。

(2) 通过 FileUploadID.PostedFile.ContentLength 可以获得通过 FileUpload 控件选择的文件大小。

(3) 通过 FileUploadID.PostedFile.SaveAs(URL)这个方法将通过 FileUpload 控件选择的文件上传到服务器相关的 URL 中。

(4) 通过 Server.MapPath()这个函数可以获取当前服务器的路径。

本实例通过在网页 FileUpload.aspx 中添加 FileUpload 控件实现文件的提交,把文件提交到本网站的 upload 文件夹中。如果提交成功就显示提交成功,如果失败,显示提交失败。

设计页面 FileUpload.aspx 的代码如下:

```
<div>
    文件上传:<asp:FileUpload ID="FileUpload1" runat="server" />
    <asp:Button ID="upload" runat="server" onclick="upload_Click" Text="上传" />
    <br />
    <asp:Label ID="show" runat="server"></asp:Label>
</div>
```

代码区域 FileUpload.aspx.cs 代码如下:

```
protected void upload_Click(object sender, EventArgs e)
{   if (FileUpload1.PostedFile !=null)            //检查文件是否上传
    {   string nam =FileUpload1.PostedFile.FileName;   //获取上传文件的文件名
        int i =nam.LastIndexOf(".");              //取得扩展名在文件名中的位置
        string newext =nam.Substring(i);          //取得文件扩展名
        DateTime now =DateTime.Now;
        String newname =now.DayOfYear.ToString()
            +FileUpload1.PostedFile.ContentLength.ToString();
                                                  //这样处理文件名是确保文件名不重复
```

```
    try
    {   //保存文件到 upload 文件夹中
        FileUpload1.PostedFile.SaveAs(Server.MapPath("~\\upload\\") +newname +
        newext);
        show.Text ="上传成功";
    }
    catch
    { show.Text ="上传失败";
    }
  }
}
```

最终运行的结果如图 3.21~图 3.23 所示。

图 3.21 运行时单击"浏览"按钮就弹出的"选择文件"对话框

图 3.22 选择文件后单击"上传"按钮出现
上传成功的页面

图 3.23 最终文件成功上传到
upload 文件夹中

3.11 Calendar 控件

在实际的网页使用中,时常会需要网页有日历功能,让用户可以很地查看当前日期以及通过日历选择日期。

在网页 Calendar.aspx 中加入 Calendar 控件,单击菜单自动套用格式可以选择一个系统中自带的样式,比如选择彩色型 1,运行结果如图 3.24 所示。

可以使用 Calendar 控件的 SelectedDate 属性,来显示当前选中的日期。在网页中加 label 控件,用来显示选中的日期。

设计页面 Calendar.aspx 的代码如下:

```
<div>
    <asp:Calendar ID="Calendar1" runat="server" OnSelectionChanged="Calendar1_SelectionChanged">
    </asp:Calendar>
    <asp:Label ID="show" runat="server"></asp:Label>
</div>
```

代码区域 Calendar.aspx.cs 代码如下:

```
protected void Page_Load(object sender, EventArgs e)
{   Calendar1.SelectedDate =DateTime.Today;          //获取当前日期
}
protected void Calendar1_SelectionChanged(object sender, EventArgs e)
{   show.Text =Calendar1.SelectedDate.ToString();
}
```

最终运行的结果如图 3.25 所示。

图 3.24 Calendar 控件的显示

图 3.25 选择日期后运行的结果

仔细观察运行结果可以发现,返回的日期值并不是单独的日期,而是带有时间,这是因为通过 Calendar 控件的 SelectedDate 属性返回的是一个 DateTime 类型的变量,这种类型的变量默认的返回值既包括日期也包括时间。如果希望单独地返回日期,而不需要时间的话,可以通过 ToShortDateString 方法实现。ToShortDateString 方法的功能是将一个 DateTime 类型的值转换为等效的短日期形式的字符串表现形式,代码如下:

```
protected void Calendar1_SelectionChanged(object sender, EventArgs e)
{    show.Text =Calendar1.SelectedDate.ToShortDateString().ToString();
}
```

3.12 小结

本章主要讲述了 ASP.NET 中基本控件的使用,包括基本控件的添加、基本控件的属性设置和基本控件的后台取值等。

课后思考问题

1. 在本章中讲述 FileUpload 控件时,仅仅讲述了如何上传文件,如果上传时还要检验上传的文件是否合法,上传的文件大小,如何实现?比方说只能上传图片文件,上传的大小不能大于 50KB。

2. 学习了基本控件的使用后,课后实现图 3.26 所示的注册功能。

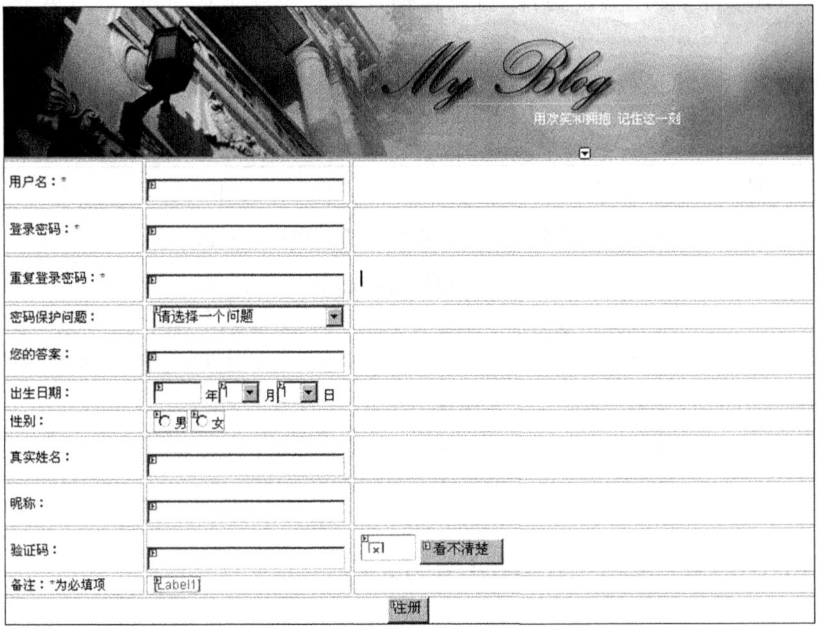

图 3.26 注册页面

第 4 章 验证控件的使用

本章要点
- 如何使用 RequiredFieldValidator 控件来确保用户提供了输入。
- 如何使用 RangeValidator 控件来保证用户的输入在某个范围内。
- 如何使用 CompareValidator 控件来进行比较。
- 如何使用 RegularExpressionValidator 控件来控制填写的内容格式正确。
- 如何使用 CustomValidator 控件。

4.1 验证控件概述

在通常的 Web 使用中,要确保用户的输入采用合适的格式,也就是说输入验证是很重要的一个方面。例如,在注册用户信息时,需要填写用户名、密码、电子邮件等信息。对于不同的用户输入信息,有着不同的验证规则。例如,用户名不能为空,密码和确认密码一致,电子邮件的输入必须符合规则等。

在 ASP.NET 中,使用验证控件,可以使输入验证变得非常简单,本章就介绍如何使用验证控件。

本章将介绍 5 种验证控件,表 4.1 总结了它们的各自的特点。

表 4.1 ASP.NET 验证控件

验证控件	验证类型	描述
RequiredFieldValidator	必填字段验证	确保在特定输入中存在数据
RangeValidator	范围验证	确保输入的数字位于两个常数之间
CompareValidator	数据类型验证和比较验证	确保输入值小于、小于等于、大于、大于等于或不等于某个常数或者用户输入值。还可以执行数据类型验证
RegularExpressionValidator	模式验证	确保字符串与指定模式匹配
CustomValidator	自定义验证	调用用户定义的功能来执行标准验证程序不能处理的验证

4.2 RequiredFieldValidator 控件

4.2.1 RequiredFieldValidator 控件的基本属性

在实际的应用中,如在用户填写表单时,有一些项目是必填项,例如用户名和密码。在 ASP.NET 中,系统提供了 RequiredFieldValidator 验证控件进行验证。使用 RequiredFieldValidator 控件能够指定某个用户在特定的控件中必须提供相应的信息,如果不填写相应的信息,RequiredFieldValidator 控件就会提示错误信息。RequiredFieldValidator 控件的主要属性如下。

(1) ID：验证控件的 ID 名称。
(2) ControlToValidate：指定验证控件用于验证的是哪个输入控件。
(3) ErrorMessage：输入错误时显示的错误信息。

这 3 个属性也是所有的验证控件都具有的基本属性。

4.2.2 RequiredFieldValidator 控件的实际使用

本实例通过在网页中添加两个文本框，分别是姓名和密码，要求这两个控件必填，用 RequiredFieldValidator 控件来实现验证。

实现步骤如下。

(1) 在页面中分别添加两个文本框，一个 ID 为 username，另一个的 ID 为 pwd，再通过工具箱中的验证部分找到 RequiredFieldValidator 控件添加两个 RequiredFieldValidator 验证控件，结果如图 4.1 所示。

(2) 设置第一个 RequiredFieldValidator 控件的 ControlToValidate 为姓名的输入文本框(username)，ErrorMessage 属性为姓名必须填写。设置第二个 RequiredFieldValidator 控件的 ControlToValidate 为密码的输入文本框(pwd)，ErrorMessage 属性为密码必须填写。结果如图 4.2 所示。

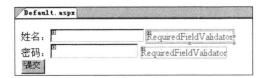
图 4.1 设计视图中的显示效果

图 4.2 添加了相应属性后的结果

(3) 完成上面两步后就能实现效果了，运行页面单击"提交"按钮，如图 4.3 所示。

图 4.3 运行结果

设计页面 Default.aspx 的代码如下：

```
<div>
    姓名:<asp:TextBox ID="username" runat="server"></asp:TextBox>
    <asp:RequiredFieldValidator ID="valName" runat="server" ErrorMessage=
    "姓名必须填写" ControlToValidate="username"></asp:RequiredFieldValidator>
    <br />
    密码:< asp:TextBox ID = "pwd" runat = "server" TextMode = "Password" ></asp:
```

```
TextBox>
<asp:RequiredFieldValidator ID="valPwd" runat="server" ErrorMessage="密码必
须填写" ControlToValidate="pwd"></asp:RequiredFieldValidator><br />
<asp:Button ID="Button1" runat="server" Text="提交" />
</div>
```

4.3 RangeValidator 控件

4.3.1 RangeValidator 控件的基本属性

范围验证控件(RangeValidator)可以检查用户的输入是否在指定的上限与下限之间。通常情况下用于检查数字、日期、货币等。范围验证控件(RangeValidator)控件的常用属性包括验证控件的基本属性和如下所示的特有属性。

（1）MinimumValue：指定有效范围的最小值。
（2）MaximumValue：指定有效范围的最大值。
（3）Type：指定要比较的值的数据类型。

4.3.2 RangeValidator 控件的实际使用

本实例通过在网页中添加一个年龄的文本框，要求这年龄控件必须在 0～200 岁之间，用 RangeValidator 控件来实现验证。

实现步骤如下。

（1）在页面中添加年龄文本框，ID 设置为 age，如图 4.4 所示。

（2）设置 RangeValidator 控件的 ControlToValidate 属性和 ErrorMessage 属性。由于 RangeValidator 控件是对 age 文本框进行验证，所以将它的 ControlToValidate 属性设置为 age，ErrorMessage 属性设置为"年龄应该在 0～200 岁之间"。

（3）然后再设置 RangeValidator 控件的 MaximumValue 和 MimimumValue 属性，MaximumValue 设置为 200，MimimumValue 设置为 0。Type 属性设置为 Integer。

（4）完成上面三步后就能实现效果了，运行页面单击"提交"按钮，如图 4.5 所示。

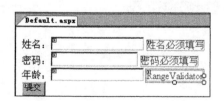

图 4.4　设计视图中的显示效果

图 4.5　运行结果

设计页面 Default.aspx 的代码如下：

年龄：`<asp:TextBox ID="age" runat="server"></asp:TextBox>`

```
<asp:RangeValidator ID="valAge" runat="server" ControlToValidate="age"
    ErrorMessage="年龄应该在 0~200 岁之间"
    MaximumValue="200" MinimumValue="0" Type="Integer"></asp:RangeValidator><br />
<asp:Button ID="Button1" runat="server" Text="提交" />
```

4.4 CompareValidator 控件

4.4.1 CompareValidator 控件的基本属性

比较验证控件对照特定的数据类型来验证用户的输入。因为当用户输入用户信息时，难免会输入错误信息，如当需要了解用户的生日时，用户很可能输入了其他的字符串。CompareValidator 比较验证控件能够比较控件中的值是否符合开发人员的需要。CompareValidator 控件的特有属性如下。

（1）ControlToCompare：以字符串形式输入的表达式。要与另一控件的值进行比较。
（2）Operator：要使用的比较。
（3）Type：要比较两个值的数据类型。
（4）ValueToCompare：以字符串形式输入的表达式。

4.4.2 CompareValidator 控件在实际中的使用

本例通过添加一个确认密码，要求确认密码跟密码一致。通过 CompareValidator 控件来实现。

（1）在页面中添加一个确认密码的文本框，并改 ID 为 repwd，添加一个 CompareValidator 控件，把它的 ControlToValidate 属性和 ErrorMessage 属性分别设置为 repwd 和"确认密码必须跟密码一致"。结果如图 4.6 所示。

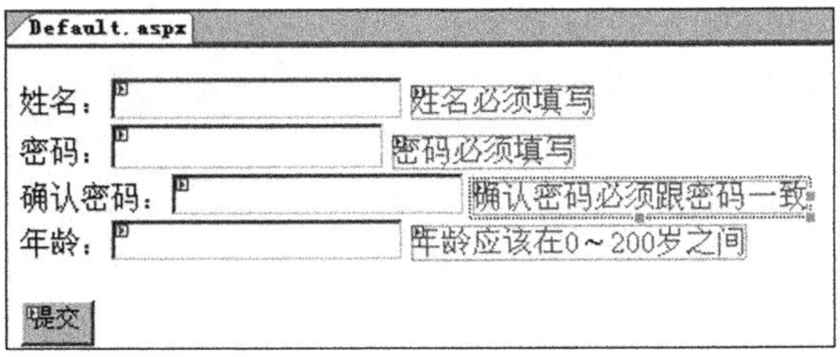

图 4.6 设计视图中添加确认密码后的结果

（2）然后设置 ControlToCompare 属性为 pwd，Type 属性为 String，Operator 属性为 Equal。
（3）最终的运行结果如图 4.7 所示。
设计页面 Default.aspx 的代码如下：

```
确认密码:<asp:TextBox ID="repwd" runat="server"></asp:TextBox>
  <asp:CompareValidator ID="valRePwd" runat="server" ControlToCompare="pwd"
ControlToValidate="repwd" ErrorMessage="确认密码必须跟密码一致" Type="String"
Operator="Equal"></asp:CompareValidator><br />
```

图 4.7 运行结果

4.5 RegularExpressionValidator 控件

4.5.1 RegularExpressionValidator 控件的基本属性

在上述控件中，虽然能够实现一些验证，但是验证的能力是有限的，例如在验证的过程中，只能验证是否是数字，或者是否是日期。也可能在验证时，只能验证一定范围内的数值，虽然这些控件提供了一些验证功能，但却限制了开发人员进行自定义验证和错误信息的开发。为实现一个验证，很可能需要多个控件同时搭配使用。

正则验证控件（RegularExpressionValidator）就解决了这个问题，正则验证控件的功能非常强大，它用于确定输入的控件的值是否与某个正则表达式所定义的模式相匹配，如电子邮件、电话号码以及序列号等。

正则验证控件（RegularExpressionValidator）常用的属性是 ValidationExpression，它用来指定用于验证的输入控件的正则表达式。客户端的正则表达式验证语法和服务端的正则表达式验证语法不同，因为在客户端使用的是 JScript 正则表达式语法，而在服务器端使用的是 Regex 类提供的正则表达式语法。使用正则表达式能够实现强大字符串的匹配并验证用户的输入的格式是否正

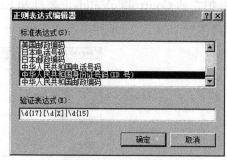

图 4.8 系统提供的正则表达式

确，系统提供了一些常用的正则表达式，开发人员能够选择相应的选项进行规则筛选，如图 4.8 所示。

4.5.2 RegularExpressionValidator 控件在实际中的使用

本例通过添加一个电子邮件，用 RegularExpressionValidator 控件验证电子邮件的地址是否合法。

（1）在页面上添加一个电子邮件的文本框，ID 为 E-mail。

（2）在页面上添加一个 RegularExpressionValidator 控件，设置它的 ControlToValidate 属性为 Email，ErrorMessage 属性为"邮件地址不正确"。

(3) 修改 RegularExpressionValidator 控件的 ValidationExpression 属性,单击 ValidationExpression 属性旁边的 按钮,就会弹出正则表达式编辑器,从中选择 Internet 电子邮件地址,最终 ValidationExpression 属性中会显示如下的正则表达式:\w+([-+.']\w+)*@\w+([-.]\w+)*\.\w+([-.]\w+)*。

(4) 最终运行的结果如图 4.9 所示。

图 4.9 加入电子邮件后的运行结果

设计页面 Default.aspx 的代码如下:

电子邮件:<asp:TextBox ID="Email" runat="server"></asp:TextBox>
<asp:RegularExpressionValidator ID="valEmail" runat="server" ControlToValidate="Email" ErrorMessage="邮件地址不正确" ValidationExpression="\w+([-+.']\w+)*@\w+([-.]\w+)*\.\w+([-.]\w+)*"></asp:RegularExpressionValidator>
<asp:Button ID="Button1" runat="server" Text="提交" />

4.6 CustomValidator 控件

自定义逻辑验证控件(CustomValidator)允许使用自定义的验证逻辑创建验证控件。例如,可以创建一个验证控件判断用户输入的是否包含"."号,示例代码如下:

```
protected void CustomValidator1_ServerValidate(object source, ServerValidateEventArgs args)
{   args.IsValid =args.Value.ToString().Contains(".");      //设置验证程序,并返回布尔值
}
protected void Button1_Click(object sender, EventArgs e)      //用户自定义验证
{
    if (Page.IsValid)                                          //判断是否验证通过
    {   Label1.Text ="验证通过";                               //输出验证通过
    }
    else
    {   Label1.Text ="输入格式错误";                           //提交失败信息
    }
}
```

上述代码不仅使用了验证控件自身的验证,也使用了用户自定义验证,运行结果如

图 4.10 所示。

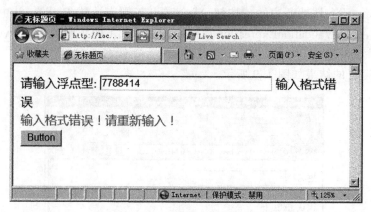

图 4.10 CustomValidator 验证控件

从 CustomValidator 验证控件的验证代码可以看出,CustomValidator 验证控件可以在服务器上执行验证检查。如果要创建服务器端的验证函数,则处理 CustomValidator 控件的 ServerValidate 事件。使用传入的 ServerValidateEventArgs 的对象的 IsValid 字段来设置是否通过验证。

而 CustomValidator 控件同样也可以在客户端实现,该验证函数可用 VBScript 或 JScript 来实现,而在 CustomValidator 控件中需要使用 ClientValidationFunction 属性指定与 CustomValidator 控件相关的客户端验证脚本的函数名称进行控件中的值的验证。

4.7 小结

本章主要讲述验证控件的使用方式,如何对于页面中的控件进行有效性验证。验证控件的使用会对用户填入的信息进行有效性验证,防止用户填写的非法数据发送到服务器端,也防止对网站的安全带来威胁。

课后思考问题

1. 做一个登录页面,在页面中加入验证控件对登录页面的输入框进行有效性校验,登录页面如图 4.11 所示。

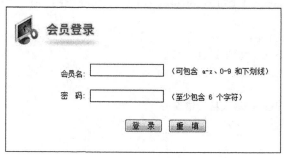

图 4.11 登录页面

2. 做一个注册页面,在页面中加入验证控件对注册页面的输入框进行有效性校验,注册页面如图 4.12 所示。

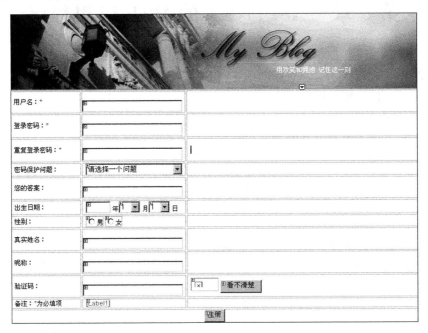

图 4.12 注册页面

第 5 章 ASP.NET 的 Web 应用程序

本章要点
- ASP.NET 应用程序基础。
- Global.asax 文件的使用。
- Application 对象的状态管理。
- Session 对象的状态管理。
- Application 与 Session 对象的使用。
- Cookie 的处理。

5.1 ASP.NET 应用程序基础

ASP.NET 应用程序(Application)是指使用 ASP.NET 技术建立的 Web 应用程序,这是由网页文件组成、可以完成特定工作的应用程序。

事实上,当在支持 ASP.NET 的 Web 服务器上建立虚拟目录后,就是在此目录中创建 ASP.NET 的 Web 应用程序。

5.1.1 网页间数据共享的基础

网页间的数据共享是 Web 应用程序能够正确执行的关键,因为 HTTP 并不会保留客户端的用户状态,用户可以通过网页间数据共享来保留用户信息,以便正确地执行 Web 应用程序。

1. Session 时间

ASP.NET 的"Session 时间"是指 Web 应用程序从一个 ASP.NET 网页跳转到其他 ASP.NET 网页过程中所花费的时间。

在 ASP.NET 中使用 System.Web.SessionState 命名空间的类来管理 Session 时间的信息,也就是建立 Session 对象来保留 Session 时间的数据。

2. 网站的数据共享

当用户进入网站,如果是会员管理网站,在登录后可以获取用户的浏览权限,但是服务器端并不知道当前客户端有哪些用户正在浏览,以及这些用户的使用状态,因为用户在移至其他网站时,用户登录数据并不会自动传给下一个网页。

不仅如此,如果 Web 网站同时有多位用户登录,就需要考虑网站的数据共享,主要分为两种。

(1) 共享给网站所有用户。在 ASP.NET 程序中可以使用 Application 变量进行共享,例如当前在线用户数。

(2) 每位用户的专用数据。在 ASP.NET 程序中可以使用 Session 变量进行共享,例如登录用户的权限。

虽然 ASP.NET 程序声明的变量无法跨越不同的 ASP.NET 程序,但是服务器端 Application 和 Session 对象建立的变量,却可以让不同的 ASP.NET 程序进行共享数据。

其目的主要是为了让用户在浏览网站的过程中,可以保留数据以便其他 ASP.NET 程序能够正常执行。

5.1.2 网页间的数据传递方法

换个角度来说,在 ASP.NET 程序间的数据传递方法就是如何保留用户状态的方法,称为"状态管理"(State Management)。按数据存储的位置分为客户端和服务器端两大类。

1. 客户端的状态管理

客户端的状态管理是将数据存储在用户计算机,或者是直接存储在 ASP.NET 程序建立的网页中,如表 5.1 所示。

表 5.1 客户端的状态管理

状态管理方法	说 明
Cookies	Cookies 是保留在用户计算机上的小文件,文件内容是一些用户信息
ViewState	ViewState 功能是在窗体回发时能够在网页中使用 ViewState 属性保留用户数据
隐藏字段	使用窗体隐藏字段回发窗体数据或传递数据到其他网页
QueryString 集合对象	使用网址 URL 参数,即在 URL 网址加上参数,将数据传递给其他网页

2. 服务器端的状态管理

服务器端的状态管理是将数据存储在服务器计算机上,换句话说,它会占用 Web 服务器的系统资源,如表 5.2 所示。

表 5.2 服务器端的状态管理

状态管理方法	说 明
Application 对象	使用 Application 对象的变量存储用户信息
Session 对象	使用 Session 对象的变量存储用户信息
数据库	使用数据库的记录存储用户信息
XML 文件或文本文件	使用 XML 文件或文本文件存储用户信息
Profile 对象	使用 HttpModules 类的 Profile 对象存储用户信息

本章主要介绍 Application 和 Session 对象的状态管理方法。

5.2 Global.asax 文件的使用

Global.asax 文件是 Web 应用程序的系统文件,属于选项文件,可有可无。当需要使用 Application 和 Session 对象的事件处理程序时,就需要创建此文件。

5.2.1 Global.asax 文件的结构

ASP.NET 的 Global.asax 文件是 ANSI 文本文件,使用 Windows 记事本就可以编辑

Global.asax 文件中的内容。

1. 创建 Global.asax 文件

Visual Studio 就可以创建 Global.asax 文件。建立网站后,右击网站名称,选择"全局应用程序类"选项(如果网站已经建立就不会看到此项目),如图 5.1 所示,单击"添加"按钮就可以在 Web 网站中建立 Global.asax 文件。一个 Web 网站只能有一个 Global.asax 文件,并且文件名不能改动。

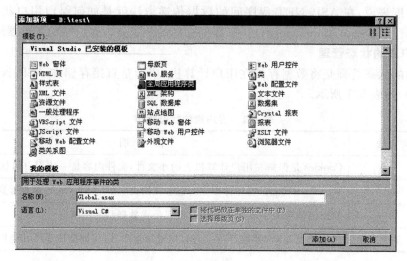

图 5.1 "添加新项"对话框

2. Global.asax 文件的结构

Global.asax 文件主要是定义 Web 应用程序的 Application_Start()、Application_End()、Session_Start()和 Session_End()等事件处理程序,文件结构如下:

```
<%@Application Language="C#" %>
<script runat="server">
void Application_Start(object sender, EventArgs e)
{    // 在应用程序启动时运行的代码
}
void Application_End(object sender, EventArgs e)
{    // 在应用程序关闭时运行的代码
}
void Application_Error(object sender, EventArgs e)
{    // 在出现未处理的错误时运行的代码
}
void Session_Start(object sender, EventArgs e)
{    // 在新会话启动时运行的代码
}
void Session_End(object sender, EventArgs e)
{    // 在会话结束时运行的代码
}
</script>
```

Application 和 Session 对象的事件处理程序如表 5.3 所示。

表 5.3 Application 和 Session 对象的事件处理程序

事件处理程序	说　　明
Application_Start()	当第 1 位用户进入 ASP.NET 程序时,Application_Start 事件就触发,触发后,就算有成千上万位用户进入网站都不会重新触发,除非 Web 服务器关机。它通常用来初始化 Application 变量,例如初始化访客的数量
Application_End()	当 Web 服务器关机时,Application_End 事件就会触发
Session_Start()	当用户建立 Session 时间时,就触发 Session_Start 事件,如果有 50 位用户,就触发 50 次事件,每个事件是独立触发的,不会相互影响,通常用来初始化用户专用的 Session 变量
Session_End()	当用户在默认时间内没有进入其他 ASP.NET 程序时,就会触发此事件,时间由 TimeOut 属性设定,通常是善后用途的程序代码,例如将 Session 变量存入数据库或文本文件

注意:当 Web.config 文件的 Sessionstate 模式设置为 InProc(此为默认值)才会触发 Session_End 事件,如果设置值为 StateServer 或 SQLServer 就不会触发此事件。

5.2.2 Global.asax 文件的使用

在 ASP.NET 的 Web 应用程序中使用 Global.asax 文件时,一个 Web 应用程序只能有唯一的 Global.asax 文件,其位置是 Web 应用程序的启动点目录。

1. Global.asax 事件处理程序的执行顺序

当用户请求 ASP.NET 程序后,就会替每位用户建立 Session 时间和 Application 对象,接着检查 ASP.NET 程序是否含有 Global.asax 文件。

如果有 Global.asax 文件,就将它编译成继承 HttpApplication 类的.NET Framework 类,然后在执行 ASP.NET 文件的程序代码前触发 Application_Start 事件,执行 Global.asax 文件的 Application_Start() 事件处理程序,并建立 Session 对象,因为 Global.asax 文件存在,接着执行 Session_Start() 事件处理程序。

当 Session 时间超过 TimeOut 属性的设定值(默认为 20min)或执行 Abandon() 方法,表示 Session 时间结束,就触发 Session_End() 事件执行处理程序,处理程序是在关闭 Session 对象前执行。

如果 Web 服务器关机,在关闭 Application 对象前就会执行 Application_End() 事件处理程序,当然也会结束所有用户的 Session 时间,执行所有用户的 Session_End() 事件处理程序。

2. ASP.NET 程序:Global.asax

在 Global.asax 文件的事件处理程序中建立程序代码,以便显示事件处理程序的执行过程,代码如下:

```
<%@Application Language="C#" %>
<script runat="server">
    void Application_Start(object sender, EventArgs e)
    {   Application["msg"] ="Application 开始…<br/>";
    }
```

```
        void Application_End(object sender, EventArgs e)
        {
        }
        void Application_Error(object sender, EventArgs e)
        {
        }
        void Session_Start(object sender, EventArgs e)
        {   Response.Write("Session 时间开始…<br/>");
            Response.Write("Applcation 变量的值为: " +Application["msg"]);
        }
        void Session_End(object sender, EventArgs e)
        {   Response.Write("Session 时间结束…<br/>");
        }
</script>
```

3. ASP.NET 程序：gloTest.aspx

在 ASP.NET 程序中显示整个 Global.asax 文件的事件处理程序的执行过程，包括网页请求事件的执行顺序，代码如下：

```
<form id="form1" runat="server">
    <div>
        显示 Global.asax 文件中的内容<br/>
    </div>
</form>
```

最终运行结果如图 5.2 所示。

图 5.2　Global.asax 文件最终运行结果

5.3　Application 对象的状态管理

在 ASP.NET 程序中可以使用 Application 和 Session 对象建立变量，可以称其为 Application 和 Session 变量，它们可以用来存储用户状态信息的共享数据。

5.3.1　Application 对象基础

Application 和 Session 对象的目的是为了保留用户状态，以便 ASP.NET 程序能够顺利执行,其差异只是保留变量的范围不同。

1. Application 对象

Application 对象可以建立 Application 变量,它和一般程序变量不同,Application 变量是一个 Contents 集合对象,此变量可以为访问网站的每一位用户提供一个共享数据的通道,因为 Application 变量允许网站的每位用户获取或更改其值。

Application 对象是在第一个 Session 对象建立后建立,Application 对象的范围直到 Web 服务器关机或所有的用户都离线后才会删除。

2. Application 对象的使用

不论网站中有多少位用户同时浏览网站,在服务器端内存中只保留一份 Application 变量,其变量格式如下:

```
Application["userid"]=1234;
```

上述程序代码将 userid 的 Application 变量设置为 1234,此变量和 ASP.NET 程序变量不一样,它是获取 Contents 集合对象的元素,变量名称是字符串。Application 对象是 System.Web 命名空间的 HttpApplicationState 类,其常用属性如表 5.4 所示。

表 5.4 Application 对象的常用属性

属性	说明
Contents	获取 HttpApplicationState 对象,此属性是为了和旧版 ASP 兼容
Count	获取共有多少个 Application 变量,也就是 HttpApplicationState 集合对象的对象数
Item	使用变量名称或索引值获取 Application 变量值

3. Application 对象状态管理的同步

ASP.NET 应用程序的每位用户都可以存取 Application 变量,用户可以同时读取 Application 变量,但是如果有一位更改数据,其他用户读取数据时,就会发生数据冲突,为了避免这种情况,需要考虑同步问题。

在 Application 对象中提供 Lock() 和 UnLock() 方法,可以保证在同一时间内只允许一位用户存取 Application 变量,其程序代码如下:

```
Application.Lock();
Application["usernum"]=Application["usernum"]+1;
Application.UnLock();
```

上述程序代码在更改变量前执行 Lock() 方法避免其他用户存取此变量,如果读取就不需要 Lock() 方法。在更改后,此例是将 Application 变量 usernum 值加 1 后,即可使用 UnLock() 方法,以便让其他用户更改此变量。Application 对象的常用方法如表 5.5 所示。

表 5.5 Application 对象的常用方法

方法	说明
Lock()	禁止其他用户修改 Application 变量
UnLock()	允许其他用户修改 Application 变量

在实际操作中，Application 变量可以记录登录网站的用户总数，或网站已经登录的用户总数，至于每位用户状态则使用 Session 变量记录。

5.3.2 网站的访客计数

网站的访客计数是一种必备组件，其目的是显示有多少位访客曾经浏览网站，显示信息可以从开站以来的访客数，或一段时间内的访客数。

ASP.NET 的访客计数在 Global.asax 文件的 Session_Start()事件处理程序中，使用 Application 变量记录访客计数，具体代码如下：

1. ASP.NET 程序：Global.asax

```
<%@Application Language="C#" %>
<script runat="server">
    void Application_Start(object sender, EventArgs e)
    {   //初始化访客计数的 Application 变量
        Application["visitedNum"] = 0;
    }
    void Session_Start(object sender, EventArgs e)
    {   Application.Lock();
        //来了一位用户计数加 1
        Application["visitedNum"] = Application["visitedNum"] + 1;
        Application.UnLock();
    }
</script>
```

2. ASP.NET 程序：showCount.aspx

```
<div>
<asp:Label ID="Label1" runat="server" Text=""></asp:Label>
</div>
```

3. ASP.NET 程序：showCount.aspx.cs

```
protected void Page_Load(object sender, EventArgs e)
{   Label1.Text = "当前在线用户为：" + Application["visitedNum"].ToString() + "位。";
}
```

最终的运行结果如图 5.3 所示。

图 5.3 显示在线用户信息

5.4 Session 对象的状态管理

对于 Web 应用程序的用户数据,或是购物车购买商品等个人专用数据,并不是使用 Application 对象,而是使用 Session 对象的状态管理。

5.4.1 Session 对象的基础

当尚未建立 Session 时间的用户浏览 Web 网站后,ASP.NET 的 Web 应用程序就会为用户建立 Session 时间和建立对应的 Session 对象。

1. Session 对象

每一个 Session 对象具有唯一的 Session ID 编号,在整个浏览 ASP.NET 应用程序的过程中(访问不同 ASP.NET 程序时),都可以存取 Session 对象建立的变量。

在 ASP.NET 的 Web 应用程序中使用 Session ID 编号判断用户是否仍在 Session 时间,它是直接到 Session 对象的 TimeOut 属性设定时间到时(默认值为 20min),或执行 Abandon()方法后才会结束 Session 时间。不过,用户每次执行新的 ASP.NET 程序时,TimeOut 属性都会归零重新计算,所以除非有浏览网站超过 TimeOut 属性的时间,否则用户浏览网站时间都属于同一个 Session 时间。如果同时有多位用户浏览网站,每位用户都指定不同的 Session ID 编号,存储此 ID 的 Session 变量只允许拥有此 ID 的用户存取,其他用户并无法存取这些变量。ASP.NET 的 Web 应用程序在同一段时间内的 Session ID 编号是唯一值,并不会重复,但是不能使用 Session ID 作为数据库表主索引,因为在不同时间,用户可能指定相同的 Session ID。

2. Session 变量的使用

Session 变量是用户的专用数据,虽然每位用户的 Session 变量名称相同,但是值可能不同。而且只有该用户才能存取自己的 Session 变量。例如,用户 hueyan 登录网站,建立 Session 变量程序代码如下:

```
Session["username"]="hueyan"
Session["password"]="1234"
```

上述程序代码的 Session 变量使用字符串作为名,username 和 password 的值属于用户 hueyan。接着另一位用户 jane 也登录网站,也会替他建立一组 Session 变量,其程序代码如下:

```
Session["username"]="jane"
Session["password"]="4567"
```

上述程序代码的 Session 变量拥有相同名称,只是值不同,因为属于不同用户的 Session 变量。

3. Session 对象的属性和方法

Session 对象的常用方法如表 5.6 所示。
Session 对象的常用属性如表 5.7 所示。

表 5.6 Session 对象的常用方法

方法	说明
Abandon()	清除用户建立的 Session 变量，也就是说再也不能存取 Session 变量的值
Remove()	删除指定的 Session 变量，参数是 Session 变量的名称字符串

表 5.7 Session 对象的常用属性

属性	说明
TimeOut	设定和获取超过 Session 时间的时间，从第 1 次进入 ASP.NET 程序到下一次请求的间隔时间，以分钟计，默认值为 20min
SessionID	获取用户唯一的 Session 编号，此为只读属性
LCID	使用指定区域码的设定，包含日期时间和货币等格式

5.4.2 目前有多少人仍在线

访客计数是历史记录，在线用户数是目前的实时状态，如果想知道目前有多少位用户停留在网站，其最大问题是如何判断用户目前仍在线，可以指定 TimeOut 属性以停留在网站多久时间来判断是否为在线用户，代码如下：

```
Session.TimeOut=5
```

上述程序代码指定值是分钟，也就说停留 5min 的用户算是目前在线用户，不过，Application_Start()事件处理程序并不能指定 TimeOut 属性，它位于 Session_Start()事件处理程序。

注释：实际上，判断用户是否仍在线是一件困难的工作，最简单的方法是使用 TimeOut 属性，但是并不能准确计算出真正仍在线的用户数，因为 TimeOut 属性值的时间差可能相差数分钟之久。

1. ASP.NET 程序：Global.asax

```
<script runat="server">
void Application_Start(object sender, EventArgs e)
{   //初始化访客计数的 Application 变量
    Application["visitedNum"] = 0;
}
    void Session_Start(object sender, EventArgs e)
{       Session.Timeout = 5;     //设定 Session 时间是多少分钟过期
    Application.Lock();
    //来了一位用户计数加 1
    Application["visitedNum"] = (Int32)Application["visitedNum"] +1;
    Application.UnLock();
}
    void Session_End(object sender, EventArgs e)
{       Application.Lock();
    //走了一位用户计数减 1
```

```
        Application["visitedNum"] = (Int32)Application["visitedNum"] -1;
        Application.UnLock();
    }
</script>
```

2. ASP.NET 程序：showCount.aspx

```
<div>
<asp:Label ID="Label1" runat="server" Text=""></asp:Label>
</div>
```

3. ASP.NET 程序：showCount.aspx.cs

```
protected void Page_Load(object sender, EventArgs e)
{   Label1.Text ="当前在线用户为: " +Application["visitedNum"].ToString() +"位。";
}
```

最终的运行结果如图 5.4 所示。

图 5.4　显示在线用户信息

5.5　Cookie 的处理

在使用 ASP.NET 建立 Web 应用程序时，要保留用户的浏览记录，例如记录用户是否浏览过网站或输入的相关数据，因为 HTTP 并不会保留状态，Cookie 就是一种存储信息的好方法。

5.5.1　Cookie 基础

Cookie 可以解决 HTTP 无法保留信息的问题。虽然用户可以使用文本文件、XML 文件和数据库来存储相关数据，但是对于成千上万个来一次或多次的访客而言，存储这些数据实在太浪费硬盘空间，Cookie 则是最佳的解决方案。

1. Cookie 存储的文件夹

Cookie 存储在客户端，也就是浏览程序所在的计算机，所以并不会浪费服务器资源，只要用户进入网站时，就可以检查客户端是否有存储 Cookie，通过 Cookie 存储信息来建立复杂的 Web 应用程序。

如果读者常常浏览 Web 网站，在 Windows XP 的 C:\Documents and Settings\Administrator\Cookie 文件夹（Administrator 是当前用户名称）中可以看到网站保留的

Cookie 文件。

2. Cookie 的使用

Cookie 在网站应用领域相当广泛，其主要目的是使用 Cookie 来保留一些数据，如下所示。

（1）个人信息。Cookie 可以保留个人信息，例如姓名、地址、时区、账号和是否曾经访问网站。

（2）个性化的信息。Cookie 可以建立个性化网站外观和个人喜好的网站内容，或者提供用户有兴趣的信息。

（3）网站购物车。在线购物车需要存储用户选购的商品，Cookie 也可以用来记录选购商品。

3. Cookie 的问题

虽然 Cookie 可以帮助用户建立复杂的 Web 应用程序，但是在使用时仍然有一些注意事项，如下所示。

（1）浏览程序是否支持 Cookie。虽然 Internet Explorer 和 Netscape 浏览程序都支持 Cookie，不过在 Internet Explorer 4.0 前的版本需要设置接受 Cookie，Internet Explorer 5.0 之后的版本默认接受 Cookie 且用户无法更改，不过，仍然有少数用户的浏览程序并不支持 Cookie。

（2）Cookie 可以删除。因为 Cookie 是存储在客户端的计算机上，所以可能在上级浏览程序或计算机软件时破坏或删除 Cookie 文件。

（3）Cookie 可能修改。因为 Cookie 存储在客户端的计算机上，用户可以自己修改 Cookie 的内容，换句话说，不可以使用 Cookie 存储重要信息。

（4）Cookie 能够复制。Cookie 文件可以复制，当然也有可能被其他用户复制。

5.5.2 如何使用 Cookie

ASP.NET 中通过 HttpCookie 类来设置或读取 Cookie 的值。Cookie 值的设置如下：

```
HttpCookie cookie =new HttpCookie("aspcn");
```

上面的代码建立起一个名为 aspcn 的 HttpCookie 实例。

建立实例后，将给其赋值。在一个 Cookie 中可以存储一个值，也可以存储多个值。通过设置 Cookie 的 Value 属性值，可以在 Cookie 中存储一个值，代码如下：

```
Cookie.Value ="deidao";
```

通过 Cookie 的 Values 集合，可以在同一个 Cookie 中存储多个值，代码如下：

```
HttpCookie cookie=new HttpCookie("aspcn");
Cookie.Values.Add("username","feidao");
Cookie.Values.Add("password","123");
```

Values 集合使用的 Add 方法中第一个参数为关键字（Key），第二个参数是设置的值（Value）。

设置 Cookie 的最后一步是通过 AppendCookie 方法将设置的 Cookie 应用到 Response 对象中去，代码如下：

```
Response.AppendCookie(cookie);
```

至此,一个简单的 Cookie 设置完毕。但是,Cookie 的功能不仅仅只是这么简单,除了设置其值外,还可以设置其应用的域(Domain)、路径(Path)、过期时间(Expires)和安全(Secure)等。

设置完 Cookie,下一步是如何读取 Cookie。

读取 Cookie,首先需要通过 Request 对象 Cookies 集合获取目标 Cookie 的对象实例。

```
HttpCookie readCookie=Request.Cookies("aspcn");
```

然后再根据 Cookie 中存储值的多少来区分读取,读取只存储一个值的 Cookie,通过读取对象 Cookie 的 Value 值即可,而读取存储有多少值的 Cookie 则读取 Values 集合中的关键字所对应的值。下面有两个例子分别演示 Cookie 存储、读取一个值和多个值的方法,并将过期时间设置为 6min 以后。

Cookie 中只存储一个值的例子如下。

1. ASP. NET 程序:cookieTest. aspx

```
<div>
<b>演示设置一个 Cookie 的值</b><br/>
单个 Cookie 的值是:<asp:Label ID="show" runat="server" Text=""></asp:Label>
</div>
```

2. ASP. NET 程序:cookieTest. aspx. cs

```
protected void Page_Load(object sender, EventArgs e)
{   HttpCookie cookie =new HttpCookie("aspcn");
    DateTime dt =DateTime.Now;
    TimeSpan ts =new TimeSpan(0, 0, 6, 0);
    cookie.Expires =dt.Add(ts);
    cookie.Value ="feidao";
    Response.AppendCookie(cookie);
    //读取 Cookie
    HttpCookie readCookie =Request.Cookies["aspcn"];
    show.Text =readCookie.Value;
}
```

最终的运行结果如图 5.5 所示。

图 5.5 Cookie 设置单个值的运行结果

Cookie 存储多个值的例子如下。

1. ASP.NET 程序：cookieTest.aspx

```
<div>
<b>演示设置多个 Cookie 的值</b><br/>
Cookie1 的值为：<asp:Label ID="show1" runat="server" Text=""></asp:Label>
Cookie2 的值为：<asp:Label ID="show2" runat="server" Text=""></asp:Label>
</div>
```

2. ASP.NET 程序：cookieTest.aspx.cs

```
protected void Page_Load(object sender, EventArgs e)
{   HttpCookie cookie =new HttpCookie("aspcn");
    DateTime dt =DateTime.Now;
    TimeSpan ts =new TimeSpan(0, 0, 6, 0);
    cookie.Expires =dt.Add(ts);
    //cookie.Value ="feidao";
    cookie.Values.Add("username", "feidao");
    cookie.Values.Add("password", "123");
    Response.AppendCookie(cookie);
    //读取 Cookie
    HttpCookie readCookie =Request.Cookies["aspcn"];
    show1.Text =readCookie.Values["username"].ToString();
    show2.Text =readCookie.Values["password"].ToString();
}
```

最终的运行结果如图 5.6 所示。

图 5.6 Cookie 存储多个值的运行结果

注意这两个程序中都有如下代码：

```
DateTime dt =DateTime.Now;
TimeSpan ts =new TimeSpan(0, 0, 6, 0);
cookie.Expires =dt.Add(ts);
```

这段代码是设置过期时间最常用的，TimeSpan 类构造函数中 4 个参数，分别代表天(Day)、小时(Hour)、分钟(Minute)和秒(Second)。最后使用 Add 方法产生新的时间，然后将其值赋给 Expires 属性。

5.6 小结

本章主要介绍了 Application 对象和 Session 对象的使用方法以及两者之间的不同点,并且把两者使用的场合做了具体介绍,本章还介绍了 Cookie 对象的使用,以及何时使用 Cookie 比较合适等方面的知识。

课后思考问题

1. Application 对象与 Session 对象的区别是什么？如何使用这两个对象？
2. Global.asax 页面的作用是什么？

第6章 ADO.NET 的应用

本章要点
- 学习和使用.NET 数据提供程序。
- 学习 Connection 连接对象的使用。
- 掌握如何打开和关闭数据库连接。
- 掌握 Command 命令对象的使用。
- 使用 DataReader 读取数据。
- 使用 DataSet 和 DataAdapter 对象查询数据。

6.1 ADO.NET 对象模型

ADO.NET 是微软公司.NET 框架的一部分,ADO.NET 类库位于 System.Data 命名空间下,由一组工具类组成,应用程序可以很轻松地借此与基于文件或基于服务器的数据存储进行通信和管理。

ADO.NET 访问数据的操作是建立到数据源的连接、执行命令以及获取返回的结果。同时,ADO.NET 提供不同的对象完成以上各步操作,如图 6.1 所示。

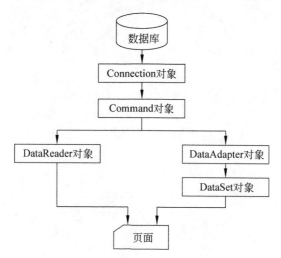

图 6.1 ADO.NET 提供多种对象

图 6.1 中涉及的 ADO.NET 的基本对象如表 6.1 所示。

表 6.1 ADO.NET 基本对象描述

对　象	描　述
Connection 对象	用来实现与数据库的连接
Command 对象	用来对数据库执行 SQL 语句

续表

对象	描述
DataReader 对象	从数据库中读取数据,实现对数据源的只读访问
DataAdapter 对象	把来自特定数据源的信息调整为关系型数据格式,以适应 DataSet 的需要
DataSet 对象	ADO.NET 的断开式组件

6.2 Connection 对象

在代码中要想对数据库操作就必须用到 Connection 对象,进行对数据库的连接。Connection 对象用于连接到数据库和管理对数据库的事务。

.NET 框架支持多种数据提供程序:根据 .NET Framework 提供程序的不同,Connection 对象分为 4 种,分别是 OleDbConnection、SqlConnection、OdbcConnection 和 OracleConnection。相应的其他对象也分别包括 4 种不同前缀的对象。

OleDbConnection 对象使用 OLEDB,任何 OLEDB 提供程序都能使用 OLEDB,包括 SQL Server。但是 SqlConnection 对象只能被 SQL Server 使用。这里主要介绍 SqlConnection 类,其他的用法跟 SqlConnection 类类似。

(1) 在使用 ADO.NET 对象的时候在.cs 文件中先把命名空间导入。使用 OLEDB 接口时导入 System.Data.OleDB 命名空间,使用 SQL Server 时导入 System.Data.SqlClient 命名空间。

(2) 把跟数据库连接的连接字符串写好。下面分别是连接 Access 数据库和 SQL Server 数据库的连接字符串。

连接 Access 数据库:

```
string strConn ="Provider=Microsoft.Jet.OLEDB.4.0;Data Source=" +Server.MapPath
("~\\App_Data\\userinfo.mdb ");
```

连接 SQL Server 数据库:

```
string strConn =" server=(local);Database=login;uid=testSa;pwd=123;";
```

(3) 连接字符串建好后,建立连接对象:

```
SqlConnection myConn =new SqlConnection(strConn);
```

(4) 建立了连接对象,使用 Open 方法把连接对象打开,就建立好了跟数据库的连接。

```
myConn.Open();
```

(5) 在使用完连接对象时,用 Close 方法把连接对象关闭。

建立连接对象的具体代码如下:

```
using System.Data.SqlClient;              //命名空间的引入
public partial class testConn : System.Web.UI.Page
{   protected void Page_Load(object sender, EventArgs e)
```

```
    {   //连接字符串设置
        string strConn ="server=(local);Database=task;uid=test;pwd=123";
        //建立连接对象
        SqlConnection myConn =new SqlConnection(strConn);
        try
        {   myConn.Open();      //用 Open 方法打开跟数据库的连接
            Response.Write("数据库连接成功!");
        }
        catch (Exception ex)
        {   Response.Write("数据库连接失败!");
            Response.Write(ex.Message);
        }
        myConn.Close();          //使用完毕用 Close 方法关闭跟数据库的连接
    }
}
```

上述是典型的数据库连接代码,对于存放数据库的连接信息,还可以有另外的两种方法。

方法一:
将数据库连接字符串存放在应用程序的配置文件(web.Config)中,代码如下:

```
<connectionStrings>
    < add name =" strConn1" connectionString =" Data Source =.\SQLEXPRESS; Initial
    Catalog= task; User  ID = test; Password = 123;" providerName =" System. Data.
    SqlClient" />
</connectionStrings>
```

这样在.cs 文件中取字符串改写如下:

```
string constr =System.Configuration.ConfigurationManager.ConnectionStrings
["strConn1"].ConnectionString;
```

方法二:
将数据库连接字符串存放在新建的一个类的方法中,例如可将此类命名为 MyClass,并在这个类中编写如下代码:

```
public class MyClass
{private static string strSql ="Data Source=.\SQLEXPRESS;Initial Catalog=task;
User ID=test;Password=123;";
    public string strCon
    {
        get{ return (strSql); }
    }
}
```

在引用此数据库连接信息时,首先要在应用程序中创建一个新类,然后再调用这个类中的属性方法。

```
MyClass class1 = new MyClass();
SqlConnection conn = new SqlConnection(class1.strCon);
conn.Open();
conn.Close();
```

以上两种方法连接字符串的重用率高,并且对于系统以后的维护带来方便。建议使用后两种方法来建立跟数据库的连接。

6.3 Command 对象

6.3.1 创建 Command 对象

在连接好数据源后,就可以对数据源执行一些命令操作。命令操作包括从数据库中检索或者对数据库进行插入、更新、删除操作。在 ADO.NET 中,对数据库的命令操作通过 Command 对象来实现。

创建 Command 对象的方法:

```
SqlCommand command = new SqlCommand(strSQL, conn);
```

参数代表的意义如下。
(1) strSQL:可以是 SQL 语句也可以是存储过程。
(2) Conn 是创建好的连接对象。
也可以这样创建 Command 对象:

```
SqlCommand comm = new SqlCommand();
comm.Connection = conn;            //指定 Command 对象的连接对象是 conn
comm.CommandText = strSQL;         //指定 Command 对象的执行语句是 strSQL
```

这两种方法起到的作用一样,都是来创建 Command 对象。

6.3.2 执行 Command 对象

创建好 Command 对象就要使用方法来执行 SQL 语句或者存储过程。Command 对象包括 3 个执行方法,如表 6.2 所示。

表 6.2 Command 对象常用方法及说明

方　法	说　明
ExecuteNonQuery	执行 SQL 语句并返回受影响的行数,使用在 insert、update、delete 语句中,返回值为 −1 说明操作不成功
ExecuteScalar	执行查询,并返回查询所返回的结果集中第一行的第一列。忽略其他列或行,在使用 count、sum、avg 等统计的时候用得多
ExecuteReader	执行返回数据集的 select 语句

6.3.3 Command 对象实现添加功能

本案例是向数据库 students 的 userInfo 表中添加一条新记录,数据库中 userInfo 的表结构如下。

(1) id:用户的序列号,表的主键,自动增长类型。
(2) username:用户名,字符串类型。
(3) pwd:密码,字符串类型。

在网站中添加一个 user_add.aspx 页面,页面上添加两个文本框,它们的 ID 分别是 username 和 pwd,最终显示如图 6.2 所示。

图 6.2 添加信息的主页面

user_add.apx 页面的代码如下:

```
<table>
    <tr><td>用户名:</td>
        <td><asp:TextBox ID="username" runat="server"></asp:TextBox></td>
    </tr>
    <tr><td>密码:</td>
        <td>< asp:TextBox ID="pwd" runat="server" TextMode="Password"></asp:
        TextBox></td>
    </tr>
    <tr><td><asp:Button ID="Button1" runat="server" Text="添加" onclick=
        "Button1_Click" /></td>
        <td><asp:Label ID="Label1" runat="server" Text=""></asp:Label></td>
    </tr>
</table>
```

user_add.aspx.cs 页面的代码如下:

```
using System.Data.SqlClient;        //注意要把命名空间加入
public partial class user_add : System.Web.UI.Page
{   protected void Page_Load(object sender, EventArgs e)
    {
    }
    protected void Button1_Click(object sender, EventArgs e)
    {   string username1 =username.Text;
```

```
        string pwd1 = pwd.Text;
        string strConn = @"server=.\SQLEXPRESS;Database=students;uid=test;pwd=123";
        SqlConnection con = new SqlConnection(strConn);
        con.Open();                             //打开数据库的连接
        string strAdd = @"insert into userInfo (username,pwd) values('" + username1 +
"','" + pwd1 + "')";
        SqlCommand com = new SqlCommand(strAdd,con);
        int i = com.ExecuteNonQuery();
        if (i > 0)
        {
            Label1.Text = "添加成功!";
        }
        else
        {
            Label1.Text = "添加失败!";
        }
        con.Close();
    }
}
```

在这个例子中,执行 insert into 语句时,使用的是 Command 对象的 ExecuteNonQuery 方法,当 ExecuteNonQuery 方法执行完后,会返回一个影响行数的整型变量,可以根据返回的整型变量来检验是否添加成功。

6.3.4 参数查询

正如 6.3.3 节讲述的插入数据的时候,用到了 insert 语句"insert into userInfo (username,pwd) values('" + username1+"','" + pwd1 + "')",给 username 和 pwd 这两列赋值时,values 后面的语句非常烦琐,并且这样写也很不安全,有 SQL 注入的危险。为了解决这些问题可以使用参数查询。

user_add.aspx.cs 页面的代码修改如下:

```
protected void Button1_Click(object sender, EventArgs e)
{   string username1 = username.Text;
    string pwd1 = pwd.Text;
    string strConn = @"server=.\SQLEXPRESS;Database=students;uid=test;pwd=123";
    SqlConnection con = new SqlConnection(strConn);
    con.Open();                             //打开数据库的连接
    string strSql = @"insert into userInfo (username,pwd) values(@name,@pwd)";
                                            //@name 代表参数查询
    SqlCommand com = new SqlCommand(strSql, con);    //创建 command 对象
    com.Parameters.Add("@name",SqlDbType.NVarChar, 50);
                                            //把@name 添加到 Command 对象的参数集合中
    com.Parameters.Add("@pwd", SqlDbType.NVarChar, 50);
                                            //把@pwd 添加到 Command 对象的参数集合中
    com.Parameters["@name"].Value = username1;       //给参数赋值
```

```
        com.Parameters["@pwd"].Value =pwd1;            //给参数赋值
        int i =com.ExecuteNonQuery();
        if (i >0)
        {   Label1.Text ="添加成功!";
        }
        else
        {   Label1.Text ="添加失败!";
        }
        con.Close();
}
```

以上代码的解释如下。

1. 参数的 SQL 命令字符串

使用参数查询的时候,只需要把 SQL 语句改写为 insert into userInfo (username,pwd) values(@name,@pwd)即可。这是以@符合开头的字符串名称。

2. 建立 Parameter 参数对象

Parameter 参数需要用 Command 对象中的 Parameter 对象的 Add()方法添加到 Parameter 集合中。代码如下:

```
com.Parameters.Add("@name",SqlDbType.NVarChar, 50);
```

第一个参数"@name"是参数名称,第二个参数 SqlDbType.NVarChar 是数据类型,第三个参数 50 是字段大小。

Paramter 参数的数据类型属于 OleDbType 或 SqlDbType 类,可以对应数据库字段的数据类型。常用数据类型如表 6.3 所示。

表 6.3　Parameter 参数的数据类型

数据类型	说　　明	对应的.NET 的数据类型
Boolean	布尔值	Boolean
Integer	32 位整数	Int32
BigInt	64 位整数	Int64
Binary	二进制数据流	Byte 数组
Currency	货币值	Decimal
Date	日期时间数据	DataTime
VarChar	不定长度的非 Unicode 字符串	String

3. 指定参数 SQL 命令的参数值

下面给每个参数指定相应的值就可以了,代码如下:

```
com.Parameters["@name"].Value =username1;      //给参数赋值
```

6.3.5　Command 对象实现更新功能

数据表中的数据如果有变更,例如用户密码改变,邮件的变更等,这就需要更新数据就行了,不需要新添加一条记录。

本案例向数据库 students 的 userInfo 表中更新一条新记录,数据库中 userInfo 的表结构如下。

(1) id：用户的序列号,表的主键,自动增长类型。
(2) username：用户名,字符串类型。
(3) pwd：密码,字符串类型。

在网站中添加一个 user_update.aspx 页面,页面上添加 3 个文本框,它们的 ID 分别是 username、pwd 和 repwd,最终显示如图 6.3 所示。

图 6.3 更新数据页面

user_update.apx 页面的代码如下：

```
<div>
    用户名：<asp:TextBox ID="username" runat="server"></asp:TextBox><br />
    密   码：<asp:TextBox ID="pwd" runat="server" TextMode="Password">
    </asp:TextBox><br />
    修改密码：<asp:TextBox ID="repwd" runat="server" TextMode="Password">
    </asp:TextBox><br />
    <asp:Button ID="Button1" runat="server" Text="确认" OnClick="Button1_Click" />
    <asp:Label ID="Label1" runat="server"></asp:Label>
</div>
```

user_update.aspx.cs 页面的代码如下：

```
protected void Button1_Click(object sender, EventArgs e)
{   //建立跟数据库的连接
    string strConn =@"Data Source=.\SQLEXPRESS;Initial Catalog=students;
    Integrated Security=True";
    SqlConnection con =new SqlConnection(strConn);
    con.Open();
    //建立 Command 对象实现 update
    string strSql ="update [userInfo] set pwd='" +repwd.Text +"' where username='"
    +username.Text +"' and pwd='" +pwd.Text +"'";
    SqlCommand com =new SqlCommand(strSql, con);
    int i =com.ExecuteNonQuery();
    if (i >0)
```

```
        {    Label1.Text ="更新数据成功!";
        }
        else
        {    Label1.Text ="更新数据失败!";
        }
        con.Close();
}
```

6.3.6　Command 对象实现删除功能

对于数据表中不再需要的记录,可以使用 SQL 的 delete 命令来删除数据表的记录数据。

本案例向数据库 students 的 userInfo 表中根据输入的用户名删除记录。

在网站中添加一个 user_delete.aspx 页面,页面上添加一个文本框,它的 ID 是 username,最终显示如图 6.4 所示。

图 6.4　删除数据页面

user_delete.apx 页面的代码如下:

```
<div>
    用户名:<asp:TextBox ID="username" runat="server"></asp:TextBox><br />
    <asp:Button ID="Button1" runat="server" OnClick="Button1_Click" Text="删除数据" OnClientClick="return confirm('你确实要删除吗?');" />
    <asp:Label ID="Label1" runat="server"></asp:Label>
</div>
```

user_delete.aspx.cs 页面的代码如下:

```
protected void Button1_Click(object sender, EventArgs e)
{    //建立跟数据库的连接
     string strConn =@"Data Source=.\SQLEXPRESS;Initial Catalog=students;Integrated Security=True";
     SqlConnection con =new SqlConnection(strConn);
     con.Open();
     //建立 Command 对象实现 delete
     string strSql ="delete from [userInfo] where username='" +username.Text +"'";
     SqlCommand com =new SqlCommand(strSql, con);
```

```
        int i =com.ExecuteNonQuery();
        if (i >0)
        {    Label1.Text ="删除数据成功!";
        }
        else
        {    Label1.Text ="删除数据失败!";
        }
        con.Close();
}
```

6.4 DataSet

6.4.1 DataSet 的数据库操作

DataSet 是 ADO.NET 的中心概念。可以把 DataSet 当成内存中的数据库,DataSet 是不依赖于数据库的独立数据集合。所谓独立,就是说,即使断开数据链路,或者关闭数据库,DataSet 依然是可用的,DataSet 在内部是用 XML 来描述数据的,由于 XML 是一种与平台无关、与语言无关的数据描述语言,而且可以描述复杂关系的数据,比如父子关系的数据,所以 DataSet 实际上可以容纳具有复杂关系的数据,而且不再依赖于数据库链路。

6.4.2 ADO.NET 的 DataSet 数据模型

因为 DataSet 可以看作是内存中的数据库,因此可以说 DataSet 是数据表的集合,它可以包含任意多个数据表(DataTable),而且每一 DataSet 中的数据表(DataTable)对应一个数据源中的数据表(Table)或是数据视图(View)。数据表实质上是由行(DataRow)和列(DataColumn)组成的集合,目的是为了保护内存中数据记录的正确性,避免并发访问时的读写冲突,DataSet 对象中的 DataTable 负责维护每一条记录,分别保存记录的初始状态和当前状态。从这里可以看出,DataSet 与只能存放单张数据表的 Recordset 是截然不同的概念。

DataSet 对象结构是非常复杂的,在 DataSet 对象的下一层中是 DataTableCollection 对象、DataRelationCollection 对象和 ExtendedProperties 对象。前面已经说过,每一个 DataSet 对象是由若干个 DataTable 对象组成。DataTableCollection 就是管理 DataSet 中的所有 DataTable 对象。表示 DataSet 中两个 DataTable 对象之间的父子关系是 DataRelation 对象。它使一个 DataTable 中的行与另一个 DataTable 中的行相关联,这种关联类似于关系数据库中数据表之间的主键列和外键列之间的关联。DataRelationCollection 对象就是管理 DataSet 中所有 DataTable 之间的 DataRelation 关系的。在 DataSet 中 DataSet、DataTable 和 DataColumn 都具有 ExtendedProperties 属性。ExtendedProperties 其实是一个属性集(PropertyCollection),用以存放各种自定义数据,如生成数据集的 select 语句等。DataRow 表示 DataType 中实际的数据,通过 DataRow 将数据添加到用 DataColumn 定义好的 DataTable。

6.4.3 DataSet 对象的三大特性

(1)独立性。DataSet 独立于各种数据源。

(2) 离线(断开)和连接。

(3) DataSet 对象是一个可以用 XML 形式表示的数据视图,是一种数据关系视图。

6.4.4 DataSet 对象的数据库操作

DataSet 对象的数据库操作需要先将数据填入 DataTable 对象,再次强调,DataSet 对象处理的对象是保存在内存的记录数据。在完成记录的插入、删除和更新后,记得使用 DataAdapter 对象更新真正的数据表记录数据。

如同 SQL 命令操作步骤,同样可以使用 Connection 对象来建立数据库连接,其建立步骤如下。

步骤一:建立 Connection 对象。

ASP.NET 程序需要使用 Connection 对象来建立数据库连接,语句如下:

```
SqlConnection con = new SqlConnection(strConn);
```

上述程序代码建立 con 的数据库连接对象。

步骤二:建立 DataAdapter 和 CommandBuilder 对象。

DataSet 对象是使用 DataAdapter 对象来获取记录数据,它是 DataSet 和 Connection 对象数据源间的桥梁,可以获取 DataSet 的记录数据和更新数据源的记录数据。

建立 DataAdapter 对象的步骤如下:

```
string strSql = "select * from userInfo";
SqlDataAdapter da = new SqlDataAdapter(strSql,con);
```

上述程序代码建立 DataAdapter 对象,第一个参数是 SQL 查询命令,第二个参数是 Connection 对象。

注意:Connection 对象一般是以 Open()方法打开数据库连接后才可以使用 DataAdapter 对象来使用它,不过建立 DataAdapter 对象可以省略 Open()方法。例如,上述程序代码就没有使用 Open()方法,而是直接使用 Connection 对象的参数来建立 DataAdapter 对象。

DataAdapter 对象提供 4 个属性,即 SelectCommand、InsertCommand、DeleteCommand 和 UpdateCommand,可以分别指定查询、插入、删除和更新 SQL 命令的 Command 对象,可以使用 CommandBuilder 对象来自动建立 SQL 命令:

```
SqlCommandBuilder cb = new SqlCommandBuilder(da);
```

上述程序代码以 DataAdapter 为参数建立 CommandBuilder 对象,此对象可以配合 DataSet 对象,当修改 DataSet 对象数据后,自动产生数据表所需的 SQL 命令,即前面所介绍的 SelectCommand、InsertCommand、DeleteCommand 和 UpdateCommand 属性值的 Command 对象。

步骤三:建立 DataSet 对象填入记录数据。

现在可以建立 DataSet 对象并且执行 DataAdapter 的 SQL 查询命令,将记录数据填入 DataSet 对象:

```
DataSet ds = new DataSet();
```

```
da.Fill(ds,"UserInfo");
```

上述程序代码建立 DataSet 对象后,使用 DataAdapter 对象的 Fill()方法将 SQL 命令的查询结果填入第一个参数的 DataSet 对象。

DataSet 对象是由多个 DataTable 对象组成,换句话说,Fill()方法可以建立 DataSet 对象的 DataTable 对象,第二个参数 UserInfo 是 DataTable 的别名,之后可以使用此名称,它是在 DataSet 对象中获取指定的 DataTable 对象。

步骤四:在 DataTable 中执行数据库操作。

DataSet 对象的每一个 DataTable 对象是对应数据库的一个数据表,当建立好 DataSet 对象填入记录数据后,就可以在 DataTable 对象中插入、删除和更新行对象 DataRow,这是一条记录。

步骤五:更新数据表的记录数据。

在 DataTable 对象中插入、更新或删除 DataRow 对象后,因为操作是针对保存在内存的 DataSet 对象,最后还需要更新数据源的记录数据,语句如下:

```
int count =da.Update(ds,"UserInfo");
```

上述程序代码使用 DataAdapter 对象的 Update()方法更新 DataSet 对象,因为已经建立 CommandBuilder 对象,在更新前,DataAdapter 对象会检查此对象,CommandBuilder 对象按照 DataTable 更新记录,自动建立 DataAdapter 对象所需的 SQL 命令来更新数据源的记录数据。

步骤六:关闭数据库连接。

最后关闭数据库连接,即使用 Connection 对象的 Close()方法关闭,语句如下:

```
con.Close();
```

6.4.5　DataSet 对象实现插入操作

DataSet 对象插入新记录操作是针对 DataTable 对象,插入记录就是新添加 DataRow 对象,语句如下:

```
DataRow newdr =ds.Tables["UserInfo"].NewRow();
```

上述程序代码使用 UserInfo 别名获取指定的 DataTable 对象后,使用 NewRow()方法新添加一行,接着可以输入字段值,语句如下:

```
newdr["username"]=username1;
newdr["pwd"] =pwd1;
```

上述程序代码指定字段值,参数字符串为字段名称,字段值是 Web 控件 TextBox 的 Text 属性值。最后将 DataRow 对象添加到 DataTable 对象,语句如下:

```
ds.Tables["userInfo"].Rows.Add(newdr);
```

上述程序代码使用 DataTable 的 Rows 属性获取 DataRowCollection 对象(DataTable 对象是每一个 DataRow 对象的集合对象),然后使用 DataRowCollection 对象的 Add()方法

将新添加的一行添加到 DataTable 的 DataRowCollection 对象。

user_add.apx 页面的代码如下：

```
<table>
    <tr>
        <td>用户名：</td>
        <td><asp:TextBox ID="username" runat="server"></asp:TextBox></td>
    </tr>
    <tr><td>密码：</td>
        <td><asp:TextBox ID="pwd" runat="server" TextMode="Password">
        </asp:TextBox></td>
    </tr>
    <tr>
        <td><asp:Button ID="Button1" runat="server" Text="添加" onclick=
        "Button1_Click" /></td>
        <td><asp:Label ID="Label1" runat="server" Text=""></asp:Label></td>
    </tr>
</table>
```

user_add.aspx.cs 页面的代码如下：

```
protected void Button1_Click(object sender, EventArgs e)
{   string username1 =username.Text;
    string pwd1 =pwd.Text;
    string strConn = @"server=.\SQLEXPRESS;Database=students;Integrated Security
    =True";
    SqlConnection con =new SqlConnection(strConn);
    con.Open();         //打开数据库的连接
    string strSql ="select * from userInfo";
    SqlDataAdapter da =new SqlDataAdapter(strSql,con);
    SqlCommandBuilder cb =new SqlCommandBuilder(da);
    DataSet ds =new DataSet();
    da.Fill(ds,"userInfo");
    //添加一行
    DataRow newdr =ds.Tables["userInfo"].NewRow();
    //填入内容
    newdr["username"] =username1;
    newdr["pwd"] =pwd1;
    //添加到 DataSet
    ds.Tables["userInfo"].Rows.Add(newdr);
    //更新数据库,也就是插入一条记录
    int i =da.Update(ds, "userInfo");
    if (i >0)
    {    Label1.Text ="添加成功！";
    }
    else
```

```
        {   Label1.Text ="添加失败!";
        }
        con.Close();
}
```

最终运行结果与6.3.3节的相似,在输入用户数据后,单击"添加"按钮,可以看到成功地插入一条记录。

6.4.6 更新记录

DataSet对象的更新记录操作,因为拥有更新条件,所以需要使用循环查找需要更新的记录,然后才能更新记录数据,代码如下:

```
foreach(DataRow dar in ds.Tables["userInfo"].Rows)
{
    if (dar["username"] ==username.Text && dar["pwd"] ==pwd.Text)
    {   dar["pwd"] =repwd.Text;
    }
}
```

上述foreach循环走访DataTable对象的所有DataRow对象,使用If条件检查是否为指定的用户账号,如果是,更新DataRow对象的密码字段。

user_update.apx页面的代码如下:

```
<div>
用户名:<asp:TextBox ID="username" runat="server"></asp:TextBox><br />
密 码:<asp:TextBox ID="pwd" runat="server" TextMode="Password">
</asp:TextBox><br />
修改密码:<asp:TextBox ID="repwd" runat="server" TextMode="Password">
</asp:TextBox><br />
<asp:Button ID="Button1" runat="server" Text="确认" OnClick="Button1_Click" />
<asp:Label ID="Label1" runat="server"></asp:Label>
</div>
```

user_ update.aspx.cs页面的代码如下:

```
protected void Button1_Click(object sender, EventArgs e)
{   //建立跟数据库的连接
    string strConn = @" Data Source =.\SQLEXPRESS; Initial Catalog = students;
    Integrated Security=True";
    SqlConnection con =new SqlConnection(strConn);
    con.Open();
    //建立Command对象实现update
    string strSql ="select * from userInfo";
    SqlDataAdapter da =new SqlDataAdapter(strSql, con);
    SqlCommandBuilder cb =new SqlCommandBuilder(da);
    DataSet ds =new DataSet();
    da.Fill(ds, "userInfo");
```

```
        //DataRow dar;
        foreach(DataRow dar in ds.Tables["userInfo"].Rows)
        {   if (dar["username"]==username.Text && dar["pwd"]==pwd.Text)
            {    dar["pwd"]=repwd.Text;
            }
        }
        int i =da.Update(ds,"userInfo");
        if (i >0)
        {    Label1.Text ="更新数据成功!";
        }
        else
        {    Label1.Text ="更新数据失败!";
        }
        con.Close();
}
```

最终运行的结果跟 6.3.5 节的相似,在输入更新用户的账号和密码后,单击"确认"按钮,可以看到更新一条记录。

6.4.7 删除记录

同样方式,可以使用 DataSet 对象来删除数据表中的记录,语句如下:

```
foreach (DataRow dar in ds.Tables["userInfo"].Rows)
{       if (dar["username"]==username.Text)
        {    dar.Delete();       //删除此记录
        }
}
```

上述 foreach 循环查看 DataTable 对象的所有 DataRow 对象,使用 If 条件检查是否是指定的用户账号,如果是,使用 Delete 方法删除 DataRow 对象。

user_delete.apx 页面的代码如下:

```
<div>
    用户名:<asp:TextBox ID="username" runat="server"></asp:TextBox><br />
    <asp:Button ID="Button1" runat="server" OnClick="Button1_Click" Text="删除数据" OnClientClick="return confirm('你确实要删除吗?');" />
    <asp:Label ID="Label1" runat="server"></asp:Label>
</div>
```

user_ delete.aspx.cs 页面的代码如下:

```
protected void Button1_Click(object sender, EventArgs e)
{   //建立跟数据库的连接
    string strConn =@"Data Source=.\SQLEXPRESS;Initial Catalog=students;Integrated Security=True";
    SqlConnection con =new SqlConnection(strConn);
    con.Open();
```

```
//建立 Command 对象实现 update
string strSql ="select * from userInfo";
SqlDataAdapter da =new SqlDataAdapter(strSql, con);
SqlCommandBuilder cb =new SqlCommandBuilder(da);
DataSet ds =new DataSet();
da.Fill(ds, "userInfo");
//DataRow dar;
foreach (DataRow dar in ds.Tables["userInfo"].Rows)
{       if (dar["username"] ==username.Text)
    {       dar.Delete();        //删除此记录
    }
}
int i =da.Update(ds, "userInfo");
if (i >0)
{    Label1.Text ="更新数据成功!";
}
else
{    Label1.Text ="更新数据失败!";
}
con.Close();
}
```

最终执行结果与6.3.6节的相似,在输入删除的用户账号后,单击"删除数据"按钮就可以看到删除一条记录。

6.5 从数据表中获取单一字段值

ADO.NET 的 Command 对象提供多个 Execute 方法,前面部分只使用了 ExecuteNonQuery()方法,本节介绍 ExecuteScalar()方法的使用,此方法可以获取数据表指定记录的单一字段值,比方说得到数据库中 Sum、Count 函数的返回值。

ExecuteScalar()方法执行 SQL 命令,如果返回值不止一个,例如多条记录,获取的是第1条记录的第1个字段。

本案例是到数据库中检索一共有多少条记录,具体代码如下。

scalarTest.aspx 页面的代码如下:

```
<div>
    userInfo 表中共有<asp:Label ID="Label1" runat="server" Text="?"></asp:
Label>条记录<asp:Button
    ID="Button1" runat="server" OnClick="Button1_Click" Text="显示" />
</div>
```

scalarTest.aspx.cs 页面的代码如下:

```
protected void Button1_Click(object sender, EventArgs e)
{    //建立跟数据库的连接
```

```
    string strConn =@"Data Source=.\SQLEXPRESS;Initial Catalog=students;
    Integrated Security=True";
    SqlConnection con =new SqlConnection(strConn);
    con.Open();
    //建立 Command 对象
    string strSql =@"select count(*) from [userInfo]";
    SqlCommand com =new SqlCommand(strSql, con);
    Label1.Text =com.ExecuteScalar().ToString();
    con.Close();
}
```

最终的运行结果如图 6.5 所示。

图 6.5　ExecuteScalar 方法的运行结果

6.6　DataReader 对象以表格显示数据表

通常使用表格在 ASP.NET 程序中显示数据表内容,以一行代表一条记录,每一列为一个字段。本节中使用表格显示数据表中的信息。第 7 章中是以数据源和 Web 控件显示数据表中的记录数据。

显示数据表中的所有数据的具体步骤如下。

(1) 建立 Connection 对象。

如同数据库操作,用户需要使用 Connection 对象来建立跟数据库的连接,代码如下:

```
string strConn = @" Data Source =.\SQLEXPRESS; Initial Catalog= students; Integrated
Security=True";
SqlConnection con =new SqlConnection(strConn);
```

上述代码建立了跟数据库的连接对象 con。

(2) 建立 Command 对象。

在建立好 Connection 对象后,就可以使用 SQL 命令和 Connection 对象为参数来建立 Command 对象,代码如下:

```
string strSql =@"select * from [userInfo]";
SqlCommand com =new SqlCommand(strSql, con);
```

1. 执行 SQL 命令查询数据表

接着使用 ExecuteReader()方法来执行 SQL 命令,可以获取 DataReader 对象,代码

如下:

```
public SqlDataReader dr;
dr = com.ExecuteReader();
```

2. 取出查询结果的数据表记录数据

DataReader 对象是一种流数据,可以使用 while 循环读取数据表的每一条记录,代码如下:

```
<%
    while(dr.Read())
    {
%>
    <tr><td><%=dr["username"].ToString()%></td><td><%=dr["pwd"].ToString()%>
        </td></tr>
<%}
    dr.Close();
%>
```

上述 while 循环使用 Read()方法读取下一条记录。也就是将记录移动到下一条,如果有下一条记录,返回 True,否则返回 False。要读取 DataReader 对象就必须用 Read()方法先看看有没有记录。

因为本例中在.aspx 页面中写入了代码,在.aspx 页面中写代码必须用<% %>来括起来。

dr["username"]是读取数据表中的字段。

3. 关闭 DataReader 和数据库连接

最后是关闭 DataReader 对象和数据库连接对象。

```
dr.Close();
con.Close();
```

show.aspx 页面的代码如下:

```
<div>
    <b>DataReader 演示</b>
    <table border="1">
        <tr bgcolor="#aaaadd"><td>用户名</td><td>密码</td></tr>
        <%    while(dr.Read())
            {
        %>
        <tr><td><%=dr["username"].ToString()%></td><td><%=dr["pwd"].ToString()%>
            </td></tr>
        <%}
            dr.Close();
        %>
    </table>
</div>
```

show.aspx.cs 页面的代码如下：

```
public SqlDataReader dr;
protected void Page_Load(object sender, EventArgs e)
{   //建立跟数据库的连接
    string strConn = @" Data Source =. \ SQLEXPRESS; Initial Catalog = students;
    Integrated Security=True";
    SqlConnection con =new SqlConnection(strConn);
    con.Open();
    string strSql =@"select * from [userInfo]";
    SqlCommand com =new SqlCommand(strSql,con);
    dr =com.ExecuteReader(CommandBehavior.CloseConnection);        //自动地关闭相应的连接对象
}
```

最终的运行结果如图 6.6 所示。

图 6.6 使用 DataReader 对象显示数据表中的记录

代码分析：

```
dr=com.ExecuteReader(CommandBehavior.CloseConnection);
```

因为如果是在 show.aspx.cs 页面中用 con.Close()的话，把连接关闭，show.aspx 页面就获取不到 dr 对象，就无法正常显示。因为 DataReader 对象是跟数据库建立的连接，在执行时加入 CommandBehavior.CloseConnection 这个参数就会用完 dr 对象后自动地关闭连接对象，就不用显式地关闭连接对象。

6.7 DataSet 对象以表格显示数据表

DataSet 对象可以建立存储在内存的数据库，它是将数据表的记录和字段数据转换为对象结构。DataSet 对象拥有多个 DataTable 对象，每一个 DataTable 对象是一个数据表，DataTable 是一个表格，可以使用行或列来存取数据，即 DataRowCollection 和 DataColumnCollection 集合对象，在集合对象的每一个 DataRow 对象是一行，即一条记录，每一个 DataColumn 对象是一列。

对于 DataTable 对象,可以建立 DataView 对象,相当于关系型数据库的视图(views),是一种定义在数据表中的虚拟数据表,简单地说,就是预先定义的数据表查询结果。

因为数据表通常是以记录为单位,所以 DataColumn 对象比较少用,通常只会使用 DataSet、DataTable、DataRowCollection 和 DataRow 对象来处理数据表。

当使用 DataAdapter 对象将数据表的记录数据填入 DataSet 对象后,就可以使用 foreach 循环从 DataRowCollection 集合对象取出每一个 DataRow 对象,代码如下:

```
foreach(DataRow dar in ds.Tables["userInfo"].Rows)
{   Response.Write("<tr>");
    Response.Write("<td>"+dar["username"]+"</td>");
    Response.Write("<td>"+dar["pwd"]+"</td>");
    Response.Write("</tr>");
}
```

show.aspx 页面的代码如下:

```
<%@Page Language="C#" AutoEventWireup="True" CodeFile="show1.aspx.cs" Inherits="show1" %>
<%@Import Namespace="System.Data" %>
<%@Import Namespace="System.Data.SqlClient" %>
<!DOCTYPE html PUBLIC "-//W3C//DTD XHTML 1.0 Transitional//EN" "http://www.w3.org/TR/xhtml1/DTD/xhtml1-transitional.dtd">
<html xmlns="http://www.w3.org/1999/xhtml" >
<head runat="server">
    <title>DataSet 显示数据表中的数据</title>
</head>
<body>
    <form id="form1" runat="server">
    <div>
    <b>DataSet 演示</b>
<table border="1">
    <tr bgcolor="#aaaadd"><td>用户名</td><td>密码</td></tr>
  <%//建立跟数据库的连接
     string strConn = @" Data Source =.\SQLEXPRESS; Initial Catalog = students;
     Integrated Security=True";
     SqlConnection con =new SqlConnection(strConn);
     con.Open();
     string strSql =@"select * from [userInfo]";
     SqlDataAdapter da =new SqlDataAdapter(strSql, con);
     DataSet ds =new DataSet();
     da.Fill(ds,"userInfo");
     foreach(DataRow dar in ds.Tables["userInfo"].Rows)
     {
        Response.Write("<tr>");
        Response.Write("<td>"+dar["username"]+"</td>");
        Response.Write("<td>"+dar["pwd"]+"</td>");
```

```
            Response.Write("</tr>");
        }
        con.Close();
    %>
</table>
    </div>
    </form>
</body>
</html>
```

最终运行的结果如图 6.7 所示。

图 6.7　使用 DataSet 对象显示数据表中的记录

6.8　小结

本章主要介绍了 ADO.NET 中包含的主要对象的使用方式,并且介绍了 ADO.NET 的两种不同的处理数据库的方式：一种是有连接的处理方式,主要使用的对象有 Connection 对象、Command 对象和 DataReader 对象等；另一种方式是无连接方式,使用的对象主要是 DataSet 和 DataAdapter、CommandBuilder 的使用。本章主要是介绍如何使用 ADO.NET 对数据库的操纵。

课后思考问题

1. 使用 Command 对象实现对数据库的增加、删除、修改功能。
2. 使用 DataReader 对象实现对数据库的查询功能。

第7章 数据控件

本章要点
- 了解数据源控件的使用。
- 掌握数据绑定知识。
- 掌握 GridView 控件的使用。
- 掌握 DataList 控件的使用。
- 掌握 Repeater 控件的使用。

在了解了 ADO.NET 基础后,就可以使用 ADO.NET 提供的对象进行数据库开发和操作。ASP.NET 还提供了一些 Web 窗体的数据控件,开发人员能够智能地配置与数据库的连接,而不需要手动编写数据库连接。ASP.NET 不仅提供了数据源控件,还提供了能够显示数据的控件,简化了数据显示的开发,开发人员只需要简单地修改模板就能够实现数据显示和分页。

7.1 数据源控件

数据源控件很像 ADO.NET 中的 Connection 对象,数据源控件用来配置数据源,当数据控件绑定数据源控件时,就能够通过数据库源控件来获取数据源中的数据并显示,而无须通过程序实现数据源代码的编写。

SqlDataSource 控件代表一个通过 ADO.NET 连接到 SQL 数据库提供者的数据源控件。并且 SqlDataSource 能够与任何一种 ADO.NET 支持的数据库进行交互,这些数据库包括 SQL Server、Access、Oledb、ODBC 以及 Oracle。

SqlDataSource 控件能够支持数据的检索、插入、更新、删除和排序等,以至于数据绑定控件可以在这些能力被允许的条件下自动地完成该功能,而不需要手动的代码实现。并且 SqlDataSource 控件所属的页面被打开时,SqlDataSource 控件能够自动地打开数据库,执行 SQL 语句或存储过程,返回选定的数据,然后关闭连接。SqlDataSource 控件强大的功能极大地简化了开发人员的开发,缩减了开发中的代码。但是 SqlDataSource 控件也有一些缺点,就是在性能上不太适应大型的开发,而对于中小型的开发,SqlDataSource 控件已经足够了。

ASP.NET 提供的 SqlDataSource 控件能够方便地添加到页面,当 SqlDataSource 控件被添加到 ASP.NET 页面中时,会生成 ASP.NET 标签,示例代码如下:

```
<asp:SqlDataSource ID="SqlDataSource1" runat="server"></asp:SqlDataSource>
```

切换到视图模式下,单击 SqlDataSource 控件会显式"配置数据源……",单击"配置数据源……"连接时,系统能够智能地提供 SqlDataSource 控件配置向导,如图 7.1 所示。

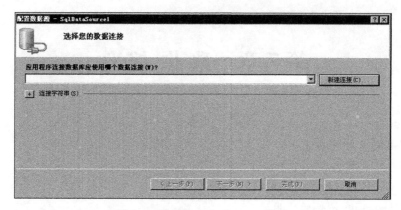

图 7.1 配置 SqlDataSource 控件

新建数据源后,开发人员可以选择是否保存在 web.config 数据源中以便应用程序进行全局配置,通常情况下选择保存。由于现在没有连接,单击"新建连接"按钮选择或创建一个数据源。单击后,系统会弹出"选择数据源"对话框,用于选择数据库文件类型,如图 7.2 所示。

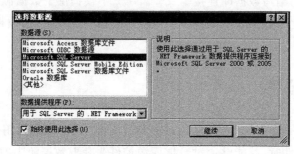

图 7.2 "选择数据源"对话框

选择数据源后,单击"继续"按钮,来配置跟数据库连接的信息,如图 7.3 所示。当配置好连接后,可以单击"测试连接"按钮来测试是否连接成功。

连接成功后,单击"确定"按钮,系统会自动添加连接,如图 7.4 所示。

单击"下一步"按钮,会出现是否将连接保存到配置文件中,如果选是的话,就会在 web.config 配置文件中建立该连接的连接字符串,当需要对用户控件进行维护时,可以直接修改 web.config,而不需要修改每个页面的数据源控件,这样就方便了开发和维护,如图 7.5 所示。

在 web.config 配置文件中添加的代码如下:

```
<connectionStrings>
    <add name="strConn" connectionString="
    Data Source = examserver03 \ SQLEXPRESS;
```

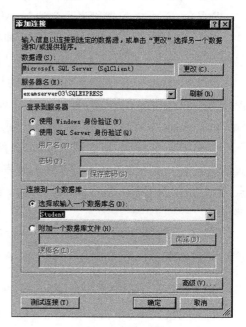

图 7.3 添加连接

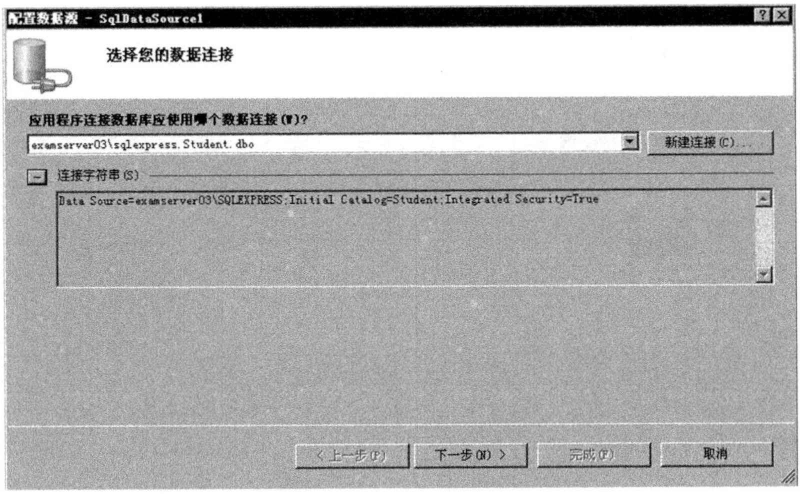

图 7.4 成功添加连接

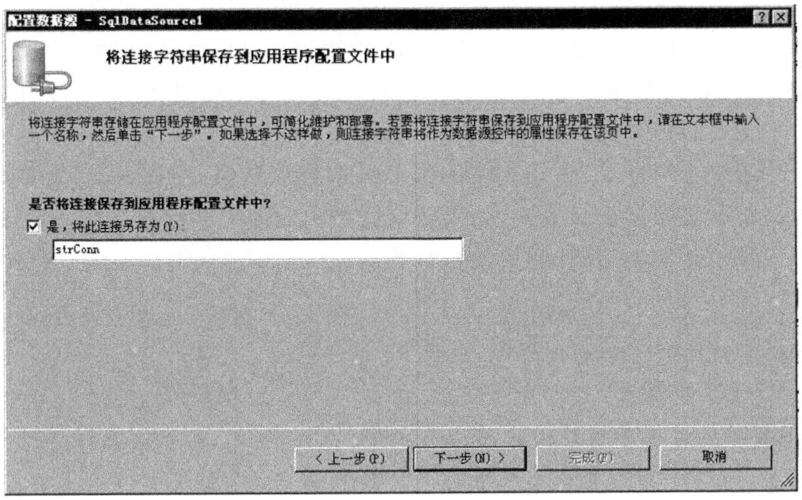

图 7.5 在配置文件中保存连接信息

　　Initial Catalog=Student;Integrated Security=True"
　　providerName="System.Data.SqlClient" />
</connectionStrings>

　　数据源控件可以指定开发人员所需要使用的 select 语句或存储过程,开发人员能够在配置 select 语句窗口中进行 select 语句的配置和生成,如果开发人员希望手动编写 select 语句或其他语句,可以单击"指定自定义 SQL 语句或存储过程"按钮进行自定义配置,select 语句的配置和生成如图 7.6 所示。

　　单击"下一步"按钮,会出现测试查询页面,单击"完成"按钮完成对数据源的配置。

　　对于开发人员,只需要勾选相应的字段,选择 Where 条件和 Order By 语句就可以配置一个 Select 语句。但是,通过选择只能够查询一个表,并实现简单的查询语。如果要实现复杂的 SQL 查询语句,可以单击"指定自定义 SQL 语句或存储过程"进行自定义 SQL 语句或

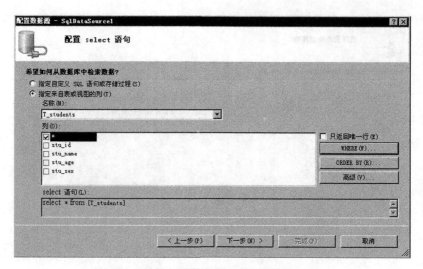

图 7.6 配置使用 select 语句

存储过程的配置,开发人员选择了一个 getdetail 的存储过程作为数据源。

单击"下一步"按钮,就需要对相应的字段进行配置,这些字段就像 ADO.NET 中的参数化查询一样。在数据源控件中,也是通过@来表示参数化变量,当需要配置相应的字段,例如配置 WHERE 语句等就需要对参数进行配置。

添加 WHERE 子句时,SQL 语句中的值可以选择默认值、控件、Cookie 或者是 Session 等。当配置完成后,就可以测试查询,如果测试后显示的结果和预期一样,则可以单击完成。

完成后,SqlDataSource 控件标签代码如下:

```
< asp:SqlDataSource ID="SqlDataSource1" runat="server" ConnectionString="<%$
ConnectionStrings:strConn %>"
    SelectCommand="select * from [T_students]"></asp:SqlDataSource>
```

注意:上例配置数据库时使用的是 Windows 身份验证,也可以选择 SQL Server 身份验证,不过 SQL Server 身份验证要求填写用户名和密码,两者产生的连接字符串不同。使用任何一种都可以。

7.2 数据绑定

数据绑定(Data Binding)的概念最早出现在 Internet Explorer 4.0,主要是使用客户端的 Dynamic HTML(DHTML)技术,ASP.NET 也支持数据绑定,能够将外部数据整合到 ASP.NET 服务器端控件。

7.2.1 数据绑定基础

ASP.NET 数据绑定是指服务器端的数据绑定,可以将外部数据整合到服务器端控件。

1. 服务器端数据绑定

服务器端数据绑定具有高扩充性、可重复性和容易维护的特点,ASP.NET 的数据绑定就是.NET Framework 的数据绑定技术,这是一种不同于 Microsoft 公司之前产品的数据绑定技术。

.NET Framework 的数据绑定技术是指将控件属性连接到任何可用数据（Data），在此的数据可以是单纯数据、对象属性、控件名称的集合对象等，.NET Framework 可以将这些数据视为类的属性来存取。

2. ASP.NET 控件与数据绑定

ASP.NET 控件支持数据绑定，可以将不同数据源的集合对象、数组、DataReader 或 DataView 对象整合到 ASP.NET 控件。

不过，不是每个 ASP.NET 控件都支持数据绑定，只有具有 DataSource 属性的控件才支持数据绑定。例如，ListBox、DropDownList、CheckBoxList、RadioButtonList、Repeater、DataList、GridView 等，才能使用。

简单地说，List 控件的数据绑定其连接数据可以视为是一维数组，也就是连接到 List 控件的选项，Repeater、DataList、GridView 等控件是二维数组的表格，可以显示整个数据表的记录数据。

在 ASP.NET 控件中使用数据绑定技术的步骤如下。

（1）定义数据源和获取数据源的数据对象。

（2）指定控件的 DataSource 属性为数据源的对象，对数据库来说就是 DataReader 或 DataView 对象。

（3）执行控件的 DataBind()方法建立数据绑定。

在执行上述步骤后，数据源的数据就会自动添入控件中，并且以控件默认方式显示数据，当然也可以设定控件属性来变更显示方式和外观。

7.2.2　ListBox 控件的数据绑定

在页面上拖放一个 ListBox 控件，不设置它的选项内容，代码如下：

```
<asp:ListBox ID="ListBox1" runat="server"></asp:ListBox>
<asp:Button ID="Button1" runat="server" onclick="Button1_Click" Text="Button" />
<asp:Label ID="Label1" runat="server" Text="Label"></asp:Label>
```

在页面上加入了一个 Button 控件和 Label 控件，当单击 Button 时在 Label 控件中显示 ListBox 控件中选中的内容。ListBox 控件中的选项是通过数据绑定的方式添加的，代码如下：

```
protected void Page_Load(object sender, EventArgs e)
{   if (!IsPostBack)      //第一次加载页面时进行下面处理
    {   string[] groups ={ ".NET框架", "JSP", "PHP", "JAVA", "XML", "HTML" };
        ListBox1.DataSource =groups;
        ListBox1.DataBind();
    }
}
protected void Button1_Click(object sender, EventArgs e)
{   Label1.Text =ListBox1.SelectedItem.Text;
}
```

最终的运行结果如图7.7所示。

图 7.7 数据绑定运行结果

7.3 数据列表控件

数据列表控件 GridView 是 ASP.NET 中功能非常丰富的控件之一,它可以以表格的形式显示数据库的内容并通过数据源控件自动绑定和显示数据。开发人员能够通过配置数据源控件对 GridView 中的数据进行选择、排序、分页、编辑和删除功能进行配置。GridView 控件还能够指定自定义样式,在没有任何数据时可以自定义无数据时的 UI 样式。

7.3.1 GridView 控件的常用事件

GridView 支持多个事件,通常对 GridView 控件进行排序、选择等操作时,同样会引发事件,当创建当前行或将当前行绑定至数据时发生的事件,同样,单击一个命令控件时也会引发事件。GridView 控件常用的事件如下。

(1) RowCommand。在 GridView 控件中单击某个按钮时发生。此事件通常用于在该控件中单击某个按钮时执行某项任务。

(2) PageIndexChanging。在单击页导航按钮时发生,但在 GridView 控件执行分页操作之前。此事件通常用于取消分页操作。

(3) PageIndexChanged。在单击页导航按钮时发生,但在 GridView 控件执行分页操作之后。此事件通常用于在用户定位到该控件中不同的页之后需要执行某项任务时。

(4) SelectedIndexChanging。在单击 GridView 控件内某一行的 Select 按钮(其 CommandName 属性设置为 Select 的按钮)时发生,但在 GridView 控件执行选择操作之前。此事件通常用于取消选择操作。

(5) SelectedIndexChanged。在单击 GridView 控件内某一行的 Select 按钮时发生,但在 GridView 控件执行选择操作之后。此事件通常用于在选择了该控件中的某行后执行某项任务。

(6) Sorting。在单击某个用于对列进行排序的超链接时发生,但在 GridView 控件执行排序操作之前。此事件通常用于取消排序操作或执行自定义的排序例程。

(7) Sorted。在单击某个用于对列进行排序的超链接时发生,但在 GridView 控件执行排序操作之后。此事件通常用于在用户单击对列进行排序的超链接之后执行某项任务。

(8) RowDataBound。在 GridView 控件中的某个行被绑定到一个数据记录时发生。此事件通常用于在某个行被绑定到数据时修改该行的内容。

(9) RowCreated。在 GridView 控件中创建新行时发生。此事件通常用于在创建某个行时修改该行的布局或外观。

(10) RowDeleting。在单击 GridView 控件内某一行的 Delete 按钮(其 CommandName 属性设置为 Delete 的按钮)时发生,但在 GridView 控件从数据源删除记录之前。此事件通常用于取消删除操作。

(11) RowDeleted。在单击 GridView 控件内某一行的 Delete 按钮时发生,但在 GridView 控件从数据源删除记录之后。此事件通常用于检查删除操作的结果。

(12) RowEditing。在单击 GridView 控件内某一行的 Edit 按钮(其 CommandName 属性设置为 Edit 的按钮)时发生,但在 GridView 控件进入编辑模式之前。此事件通常用于取消编辑操作。

(13) RowCancelingEdit。在单击 GridView 控件内某一行的 Cancel 按钮(其 CommandName 属性设置为 Cancel 的按钮)时发生,但在 GridView 控件退出编辑模式之前。此事件通常用于停止取消操作。

(14) RowUpdating。在单击 GridView 控件内某一行的 Update 按钮(其 CommandName 属性设置为 Update 的按钮)时发生,但在 GridView 控件更新记录之前。此事件通常用于取消更新操作。

(15) RowUpdated。在单击 GridView 控件内某一行的 Update 按钮时发生,但在 GridView 控件更新记录之后。此事件通常用来检查更新操作的结果。

(16) DataBound。此事件继承自 BaseDataBoundControl 控件,在 GridView 控件完成到数据源的绑定后发生。

7.3.2 使用 GridView 控件绑定数据源

下面示例先利用 SqlDataSource 控件配置数据源,连接数据后,使用 GridView 控件绑定数据源。

程序实现主要步骤如下。

(1) 新建一个网站,默认主页为 Default.aspx。添加 1 个 GridView 控件和 1 个 SqlDataSource 控件,如图 7.8 所示。

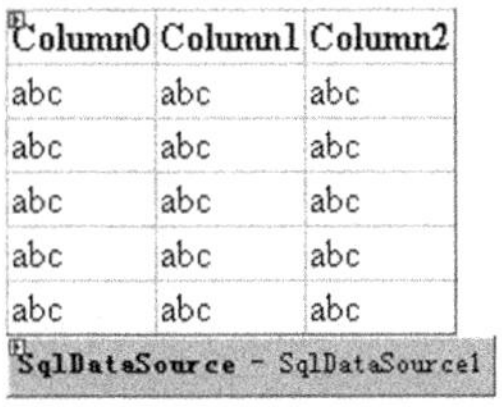

(2) 配置 SqlDataSource 控件:所有的配置请参照 7.1.1 节。

图 7.8 在页面中添加 GridView 控件和 SqlDataSource 控件后的效果

(3) 将获取的数据源绑定到 GridView 控件上,如图 7.9 所示。

(4) 单击 GridView 控件右上方的▶按钮,在弹出的快捷菜单中选择"编辑列"选项,如图 7.10 所示。

(5) 将每个 BoundField 控件绑定字段的 HeaderText 属性设置为该列头标题名,把 DataField 属性设置为字段名。"字段"对话框如图 7.11 所示。

图 7.9 绑定 GridView 控件的数据源　　　　图 7.10 选择"编辑列"选项

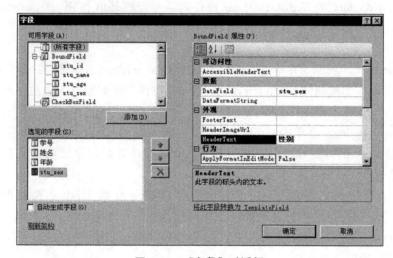

图 7.11 "字段"对话框

执行程序,示例运行结果如图 7.12 所示。

图 7.12 使用 GridView 控件绑定数据源

7.3.3 设置 GridView 控件的外观

默认状态下,GridView 控件的外观是简单的表格。为了美化网页的界面,丰富页面的显示效果,开发人员可以通过多种方式来美化 GridView 控件的外观。

GridView 控件常用外观属性如表 7.1 所示。

表 7.1　GridView 控件常用外观属性及说明

属　　性	说　　明
BackColor	用来设置 GridView 控件的背景颜色
BackImageUrl	用来设置 GridView 控件的背景中的图片的 URL
BorderColor	用来设置 GridView 控件的边框颜色
BorderStyle	用来设置 GridView 控件的边框样式
BorderWidth	用来设置 GridView 控件的边框的宽度
Caption	用来设置 GridView 控件的标题文字
CaptionAlign	用来设置 GridView 控件的标题文字的布局位置
CellPadding	用来设置 GridView 控件的单元格内容和单元格之间的空间量
CellSpacing	用来设置 GridView 控件的单元格的空间量
CssClass	用来设置 GridView 控件在客户端呈现的 CSS 样式
Font	用来设置 GridView 控件的关联的字体属性
ForeColor	用来设置 GridView 控件的前景色
GridLines	用来设置 GridView 控件的网格线样式
Height	用来设置 GridView 控件的高度
HorizontalAlign	用来设置 GridView 控件在页面上的水平对齐方式
ShowFooter	用来设置 GridView 控件是否显示页脚
ShowHeader	用来设置 GridView 控件是否显示页眉
Width	用来设置 GridView 控件的宽度

还可以通过自动套用格式来改变 GridView 控件的外观。单击"自动套用格式",弹出"自动套用格式"对话框,如图 7.13 所示。在对话框左侧可以看到 GridView 给出的几种样式。

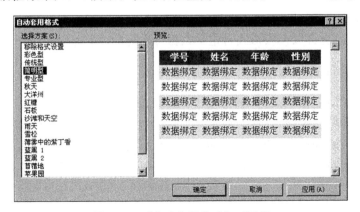

图 7.13　"自动套用格式"对话框

7.3.4 制定 GridView 控件的列

GridView 控件中的每一列由一个 DataContolField 对象表示。默认情况下，AutoGenerateColumns 属性被设置为 True，为数据源中的每一个字段创建一个 AutoGenerateColumns 对象。将 AutoGenerateColumns 属性设置为 False 时，可以自定义数据绑定列。GridView 控件共包括 7 个类型的列，分别为 BoundField（普通数据绑定列）、CheckBoxField（复选框数据绑定列）、CommandField（命令数据绑定列）和 ImageField（图片数据绑定列）、HyperLinkField（超链接数据绑定列）、ButtonField（按钮数据绑定列）、TemplateField（模板数据绑定列）。

1. BoundField

BoundField 是默认的数据绑定类型，通常用于显示普通文本。

2. CheckBoxField

使用 CheckBoxField 控件显示布尔类型的数据。绑定数据为 True 时，复选框数据绑定列为选中状态；绑定数据为 False 时，则显示未选中状态。在正常情况下，CheckBoxField 显示在表格中的复选框控件处于只读状态。只有 GridView 控件的某一行进入编辑状态后，复选框才恢复为可修改状态。

3. CommandField

CommandField 显示用来执行选择、编辑或删除操作的预定义命令按钮。这些按钮可以呈现为普通按钮、超链接和图片等外观。

4. ImageField

ImageField 在 GridView 控件呈现的表格中显示图片列。通常 ImageField 绑定的内容是图片的路径。

5. HyperLinkField

HyperLinkField 允许将所绑定的数据以超链接的形式显示出来。开发人员可自定义绑定超链接的显示文字、超链接的 URL，以及打开窗口的方式等。

6. ButtonField

ButtonField 也可以为 GridView 控件创建命令按钮。开发人员可以通过按钮来操作其所在行的数据。

7. TemplateField

TemplateField 允许以模板形式自定义数据绑定列的内容。

7.3.5 查看 GridView 控件中数据的详细信息

在设计网页显示商品信息时，通常需要考虑页面的美观程度只把概要信息显示出来。需要查看商品的详细信息时，通常会使用一个超级链接按钮链接到显示商品详细信息的页面。这样做不但使页面简化、美观，还可以提高显示的速度。下面使用 GridView 控件来显示商品的概要并使用 GridView 控件里的 HyperLinkField（超链接数据绑定列）链接到商品详细信息页做一个实例。

程序实现的主要步骤如下。

（1）新建一个网站，默认主页为 Default.aspx。添加 1 个 SqlDataSource 控件和 1 个

GridView 控件,再建立一个页面命名为 showStudent.aspx。在 showStudent.aspx 页面里添加 4 个 TextBox 控件用来显示学生的详细信息。

(2) 配置 SqlDataSource 数据源,在配置 Select 语句时选择 *（所有字段）。

(3) 在 GridView 编辑列里添加一个 HyperLinkField(超链接数据绑定列)。把超级链接列的 DataNavigateUrlFields 属性设为 stu_id,DataNavigateUrlFormatString 属性设为 showStudent.aspx?id={0},HeaderText 属性设为"查看详细信息",NavigateUrl 属性设为 ~/showStudent.aspx,Text 属性设为"详细信息",如图 7.14 所示。

图 7.14 在 GridView 控件中编辑列的设置

(4) 主要程序代码。

在 showStudent.aspx 页面里接收超链接控件传送的编号。通过编号在数据表里查询出相同编号的数据,把数据显示在 showStudent.aspx 页面里显示出来。实现这个功能需要一个 Bind()方法。Bind()方法的代码如下:

```
protected void Bind()
{   //建立数据库连接
    string strConn =@"Data Source=.\SQLEXPRESS;Initial Catalog=Student;Integrated Security=True";    //Windows 身份验证
    SqlConnection con =new SqlConnection(strConn);
    con.Open();
    //创建 Command 对象
    string strSql ="select * from T_students where stu_id='" +Request.QueryString["id"] +"'";
    SqlCommand com =new SqlCommand(strSql, con);
    //执行 SQL 语句
    SqlDataReader dr =com.ExecuteReader();

    if (dr.Read())
    {
        sid.Text =dr["stu_id"].ToString();
```

```
            sName.Text =dr["stu_name"].ToString();
            sAge.Text =dr["stu_age"].ToString();
            sSex.Text =dr["stu_sex"].ToString();
        }
        dr.Close();
        con.Close();
    }
```

注意：此方法中应用了 System.Data.SqlClient 命名空间中的 SqlConnection 对象,需要引用 System.Data.SqlClient 命名空间,引用后才能使用 SqlConnection 对象,以下实例跟这相似,将不再提起引入命名空间的问题。

在页面的 Page_Load 事件里面调用自定义方法 Bind(),代码如下：

```
protected void Page_Load(object sender, EventArgs e)
{   if (!IsPostBack)
    {   Bind();
    }
}
```

运行实例,结果如图 7.15 所示。

图 7.15　查看详细信息首页

单击王新后面的"详细信息"超链接,将显示学生的详细信息,如图 7.16 所示。

图 7.16　跳转到详细信息页

7.3.6 使用 GridView 控件分页显示数据

在网站开发过程中,经常需要表格控件查看一些基本信息,由于信息过多表格会变长,这样网页会不美观。这种情况可以使用 GridView 控件的分页功能,查看信息更为方便。GridView 控件的分页功能也非常方便,只需设置如图 7.17 所示的启用分页即可。

程序实现的主要步骤如下:新建一个网站,默认主页为 Default.aspx。添加 1 个 GridView 控件。首先,将 GridView 控件的 AllowPaging 属性设置为 True,表示允许分页。然后,将 PageSize 属性设置一个数字,用来控制每个页面中显示的记录个数,这里设为 4。再使用 SqlDataSource 数据源控件来配置数据源。最后,重新绑定 GridView 控件即可。

实例运行效果如图 7.18 所示。

图 7.17 启用分页

图 7.18 GridView 控件分页实例

7.3.7 在 GridView 控件中实现全选和全不选功能

在管理网站或在购物车中,需要对所有信息或商品进行操作时,通常是使用一个全选的 CheckBox 控件来实现这些操作。下面实例利用这个全选的 CheckBox 控件实现全选或全消的功能。

程序实现的主要步骤如下。

(1) 新建一个网站,默认主页为 Default.aspx。添加 1 个 GridView 控件和 1 个 CheckBox 控件。CheckBox 控件的 AutoPostBack 属性设为 True。代码如下:

```
<asp:CheckBox ID="CheckBox1" runat="server" AutoPostBack="True" /> 
    <asp:GridView ID="GridView1" runat="server">
    </asp:GridView>
```

(2) 添加一个 SqlDataSource 控件,配置数据源。
(3) 首先为 GridView 控件添加一列模板列,然后向模板列中添加 CheckBox 控件。GridView 控件的设计代码如下:

```
<asp:GridView ID="GridView1" runat="server">
    <Columns>
        <asp:TemplateField HeaderText="选择">
```

```
            <ItemTemplate>
                <asp:CheckBox ID="CheckBox2" runat="server" Text="选择" />
            </ItemTemplate>
        </asp:TemplateField>
    </Columns>
</asp:GridView>
```

(4) 设置 GridView 控件的数据源为 SqlDataSource1，并且把绑定字段都设置好。代码如下：

```
<asp:GridView ID="GridView1" runat="server" AutoGenerateColumns="False" DataKeyNames="stu_id" DataSourceID="SqlDataSource1">
    <Columns>
        <asp:TemplateField HeaderText="选择">
            <ItemTemplate>
                <asp:CheckBox ID="CheckBox2" runat="server" Text="选择" />
            </ItemTemplate>
        </asp:TemplateField>
        <asp:BoundField DataField="stu_id" HeaderText="学号" ReadOnly="True" SortExpression="stu_id" />
        <asp:BoundField DataField="stu_name" HeaderText="姓名" SortExpression="stu_name" />
        <asp:BoundField DataField="stu_age" HeaderText="年龄" SortExpression="stu_age" />
        <asp:BoundField DataField="stu_sex" HeaderText="性别" SortExpression="stu_sex" />
    </Columns>
</asp:GridView>
```

(5) 主要程序代码。

改变"全选"复选框的选项状态时，将循环访问 GridView 控件中的每一项，并通过 FindControl 方法搜索 TemplateField 模板列中 ID 为 CheckBox2 的 CheckBox 控件，并建立该控件的引用，实现全选/不全选功能。在 CheckBox1 控件中有 CheckedChanged 事件，在 CheckedChanged 事件中编写代码如下：

```
protected void CheckBox1_CheckedChanged(object sender, EventArgs e)
{   for (int i=0; i<=GridView1.Rows.Count -1; i++)
    {  //建立模板列中 CheckBox 控件的引用
        CheckBox cb = (CheckBox)GridView1.Rows[i].FindControl("CheckBox2");
        if (CheckBox1.Checked ==True)
        {  cb.Checked =True;
        }
        else
        {  cb.Checked =False;
        }
    }
}
```

运行实例,单击全选所有信息的 CheckBox 控件为选中状态,如图 7.19 所示。

图 7.19 GridView 控件中实现全选和全不选功能

7.3.8 在 GridView 控件中对数据进行编辑操作

在 GridView 控件的按钮列中包括"编辑"、"更新"、"取消"按钮,这 3 个按钮分别触发 GridView 控件的 RowEditing、RowUpdating、RowCancelingEdit 事件,从而完成对指定项的编辑、更新和取消操作的功能。下面实例利用 GridView 控件的 RowCancelingEdit、RowEditing 和 RowUpdating 事件,对指定项的信息进行编辑操作。

程序实现的主要步骤如下。

(1) 新建一个网站,默认主页为 Default.aspx。添加 1 个 GridView 控件和 1 个 SqlDataSource 控件。

(2) 配置 SqlDataSource 控件,在配置的时候一定要把高级部分选中,如图 7.20 所示。

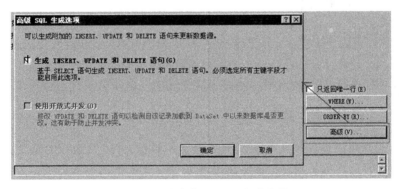

图 7.20 选中高级 SQL 生成选项

(3) 设置 GridView 控件的数据源为 SqlDataSource1,并且编辑 GridView 控件绑定字段,使结果显示如图 7.21 所示。

(4) 选中 GridView 控件,选择菜单中的"启用编辑"复选框和"启用删除"复选框,即可实现编辑和删除功能,如图 7.22 所示。

(5) 最终运行结果如图 7.23 所示。

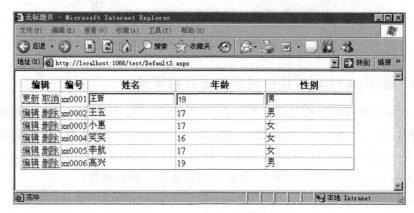

图 7.21 GridView 控件显示样式　　　　图 7.22 选中编辑和删除功能

图 7.23 GridView 控件实现编辑功能

7.4 DataList 控件

DataList 控件可以显示模板定义的数据绑定列表，其内容可以通过使用模板进行控制。通过使用 DataList 控件，用户可以显示、选择和编辑多种不同数据源中的数据。本节讨论如何使用 DataList 控件进行分页，查看数据的详细信息，使用 DataList 绑定数据源等一些功能。

7.4.1 DataList 控件概述

DataList 控件可以使用模板和定义样式来显示数据，并进行数据的选择、删除，以及编辑。DataList 控件最大的特点是一定要通过模板来定义数据的显示格式。要设计出美观的界面，就需要花费一番心思。正因为如此，DataList 控件显示数据时却更具灵活性，开发人员个人发挥的空间比较大。DataList 控件支持的模板如下。

1. AlternatingItemTemplate

如果已定义，则为 DataList 中的交替项提供内容和布局；如果未定义，则使用 ItemTemplate。

2. EditItemTemplate

如果已定义，则为 DataList 中当前编辑项提供内容和布局；如果未定义，则使用 ItemTemplate。

3. FooterTemplate

如果已定义,则为 DataList 的脚注部分提供内容和布局;如果未定义,将不显示脚注部分。

4. HeaderTemplate

如果已定义,则为 DataList 的页眉节提供内容和布局;如果未定义,将不显示页眉节。

5. ItemTemplate

ItemTemplate 为 DataList 中的项提供内容和布局所要求的模板。

6. SelectedItemTemplate

如果已定义,则为 DataList 中当前选定项提供内容和布局;如果未定义,则使用 ItemTemplate。

7. SeparatorTemplate

如果已定义,则为 DataList 中各项之间的分隔符提供内容和布局;如果未定义,将不显示分隔符。

7.4.2 DataList 控件常用的属性、方法和事件

下面对常用的属性进行详细介绍。

1. DataKeyFiled

DataKeyFiled 属性指定由 DataSource 属性指示的数据源中的键字段。键字段是数据表中唯一的字段,一般情况下使用这个字段做索引。所指定的字段用于填充 DataKeys 集合。可以用数据列表控件存储键字段而无须在控件中显示它。

2. DataKeys

使用 DataKeys 集合访问数据列表控件中每个记录的键值(显示为一行)。这样可以用数据列表控件存储键字段而无须在控件中显示它。此集合自动用 DataKeyField 属性指定的字段中的值填充。

下面对常用的方法进行详细介绍。

FindControl 方法

FindControl 方法是在当前的命名容器中搜索指定的服务器控件。该方法在 DataList 控件或 GridView 控件中会经常用到。

例如,在 DataList 控件中查找 ID 为 TextBoxName 的 TextBox 控件的 Text 属性值使用如下代码:

```
String Name =((TextBox)e.Item.FindControl("TextBoxName")).Text;
```

下面对常用的事件进行详细介绍。

(1) EditCommand 事件

对 DataList 控件中的某个项单击 Edit 按钮时发生 EditCommand 事件。

(2) UpdateCommand 事件

对 DataList 控件中的某个项单击 Update 按钮时发生 UpdateCommand 事件。

(3) DeleteCommand 事件

对 DataList 控件中的某个项单击 Delete 按钮时发生 DeleteCommand 事件。

(4) CancelCommand 事件

对 DataList 控件中的某个项单击 Cancel 按钮时发生 CancelCommand 事件。

7.4.3 使用 DataList 控件绑定数据源

DataList 控件绑定数据源的方法与 GridView 控件基本相似,但要将所绑定数据源的数据显示出来,这需要通过设计 DataList 控件的模板来完成。下面示例介绍了如何使用 DataList 控件的模板显示绑定的数据源数据。

程序实现的主要步骤如下。

(1) 新建一个网站,默认主页为 Default.aspx。添加 1 个 DataList 控件。

(2) 单击 DataList 控件右上方的 ▶ 按钮,在弹出的快捷菜单中的选择"编辑模板"选项。打开"DataList 任务—模板编辑模式",在"显示"下拉列表框中选择 HeaderTemplate 选项,如图 7.24 所示。

(3) 在 DataList 控件的页眉模板中添加一个表格用于布局和显示表头,如图 7.25 所示。

(4) 在"DataList 任务—模板编辑模式"中选择 ItemTemplate 选项,打开项模板。同样在项模板中添加一个用于布局的表格,并添加 4 个 Label 控件用于显示数据源中的数据记录,Label 控件的 ID 属性分别为 ID、Name、Sex 和 Age,如图 7.26 所示。

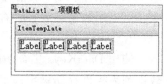

图 7.24 "DataList 任务"快捷菜单　　7.25 设计 DataList 页眉模板　　图 7.26 设计项模板

(5) 单击 ID 属性名为 ID 的 Label 控件右上角的"Label ▶"标记,打开"Label 任务"快捷菜单,执行"编辑 DataBindings"命令,打开"ID DataBindings"对话框。在 Text 属性的代码表达式对话框中写入 Eval("ID"),用于绑定数据源中的 ID 字段,如图 7.27 所示。

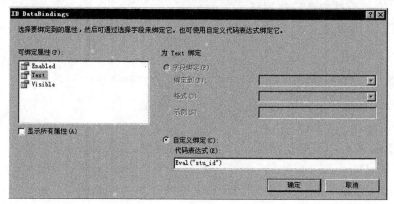

图 7.27　ID DataBindings 对话框

其他 3 个 Label 控件绑定方法同上。以上设计 DataList 控件的源代码如下：

```
<asp:DataList ID="DataList1" runat="server">
    <HeaderTemplate>
        <table>
            <tr>
                <td>学号</td>
                <td>姓名</td>
                <td>性别</td>
                <td>年龄</td>
            </tr>
        </table>
    </HeaderTemplate>
    <ItemTemplate>
        <table>
            <tr>
                <td><asp:Label ID="ID" runat="server" Text='<%#Eval("stu_id")%>'>
                </asp:Label></td>
                <td><asp:Label ID="Name" runat="server" Text='<%#Eval("stu_name") %>'></asp:Label></td>
                <td><asp:Label ID="Sex" runat="server" Text='<%#Eval("stu_sex")%>'></asp:Label></td>
                <td><asp:Label ID="Age" runat="server" Text='<%#Eval("stu_age")%>'></asp:Label></td>
            </tr>
        </table>
    </ItemTemplate>
</asp:DataList>
```

(6) 在"DataList 任务—模板编辑模式"中单击"结束模板编辑"选项，结束模板编辑。
(7) 自定义方法 Bind()，用来将控件绑定到数据源，代码如下：

```
protected void Bind()
{   //实例化 SqlConnection 对象
    string strConn =@"Data Source=.\SQLEXPRESS;
    Initial Catalog=Student;Integrated Security=True";
    SqlConnection con =new SqlConnection(strConn);
    con.Open();
    //实例化 SqlCommand 对象
    string strSql ="select * from T_students";
    SqlCommand com =new SqlCommand(strSql,con);
    SqlDataReader dr =com.ExecuteReader();
    //绑定 DataList 控件的数据源
    DataList1.DataSource =dr;
    DataList1.DataBind();
    dr.Close();
```

```
        con.Close();
    }
```

在页面的 Page_Load 事件里调用自定义方法 Bind(),代码如下:

```
protected void Page_Load(object sender, EventArgs e)
{   if (!IsPostBack)
    {  Bind();
    }
}
```

运行实例效果如图 7.28 所示。

图 7.28　DataList 控件绑定数据源

7.4.4　分页显示 DataList 控件中的数据

DataList 控件并没有类似 GridView 控件中与分页相关的属性,那么 DataList 控件是通过什么方法实现分页显示呢? 其实很简单,只要借助 PagedDataSource 类来实现就可以了,该类封装数据绑定控件与分页相关的属性,以允许该控件执行分页操作。下面实例介绍了如何使用 PagedDataSource 类实现 DataList 控件的分页功能。

程序实现的主要步骤如下。

(1) 新建一个网站,默认主页为 Default.aspx。添加 1 个 DataList 控件、2 个 Label 控件和 4 个 LinkButton 控件。Label 控件分别用于显示当前页和总页数。LinkButton 控件分别用于显示第一页、上一页、下一页、最后一页。

(2) 主要程序代码。

页面加载时调用自定义方法 Bind(),Bind()方法的具体代码如下:

```
protected void Bind()
{   //获取当前页码
    int curpage =Convert.ToInt32(this.currentPage.Text);
    //生成一个 PagedDataSource 对象
    PagedDataSource ps =new PagedDataSource();
```

```
string strConn =@"Data Source=.\SQLEXPRESS;Initial Catalog=Student;
Integrated Security=True";
SqlConnection con =new SqlConnection(strConn);
con.Open();
string strSql ="select * from T_students";
SqlCommand comm =new SqlCommand(strSql, con);
SqlDataAdapter odbA =new SqlDataAdapter(strSql, con);
DataSet ds =new DataSet();
odbA.Fill(ds, "students");
ps.DataSource =ds.Tables["students"].DefaultView;
ps.AllowPaging =True;
ps.PageSize =3;
ps.CurrentPageIndex =curpage -1;
this.first.Enabled =True;
this.front.Enabled =True;
this.next.Enabled =True;
this.last.Enabled =True;
if (curpage ==1)
{   this.first.Enabled =False;
    this.front.Enabled =False;
}
if (curpage ==ps.PageCount)
{   this.next.Enabled =False;
    this.last.Enabled =False;
}
this.totalPage.Text =Convert.ToString(ps.PageCount);
this.DataList1.DataSource =ps;
this.DataList1.DataBind();
con.Close();
}
```

在"首页"、"上一页"、"下一页"、"尾页"按钮的 Click 事件中设置当前页数,然后重新绑定数据,代码如下:

```
protected void first_Click(object sender, EventArgs e)
{
    currentPage.Text =Convert.ToString(1);
    Bind();
}
protected void front_Click(object sender, EventArgs e)
{
    currentPage.Text =Convert.ToString(Convert.ToInt32(currentPage.Text) -1);
    Bind();
}
protected void next_Click(object sender, EventArgs e)
```

```
{
    currentPage.Text =Convert.ToString(Convert.ToInt32(currentPage.Text) +1);
    Bind();
}
protected void last_Click(object sender, EventArgs e)
{
    currentPage.Text =totalPage.Text;
    Bind();
}
```

在页面的 Page_Load 事件里调用自定义方法 Bind(),代码如下：

```
protected void Page_Load(object sender, EventArgs e)
{
    if (!IsPostBack)
    {   Bind();
    }
}
```

运行实例效果如图 7.29 所示。

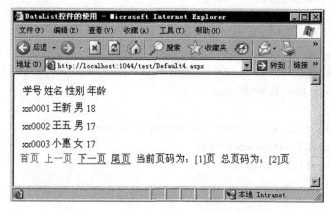

图 7.29 分页显示 DataList 控件中的数据

7.4.5 查看 DataList 控件中数据的详细信息

显示被选择记录的详细信息可以通过 SelectedItemTemplate 模板来完成。使用 SelectedItemTemplate 模板显示信息时,需要有一个控件激发 DataList 控件的 ItemCommand 事件。下面示例介绍了如何使用 SelectedItemTemplate 模板显示 DataList 控件中数据的详细信息。

程序实现的主要步骤如下。

(1) 新建一个网站,默认主页为 Default.aspx。在 Default.aspx 页中添加 1 个 DataList 控件。

(2) 打开 DataList 控件的项模板编辑模式。在 ItemTemplate 模板中添加 1 个 LinkButton 控件,用于显示用户选择的数据项;在 SelectedItemTemplate 模板中添加 1 个

LinkButton 控件和 4 个 Label 控件,分别用来取消对该数据项的选择和该数据项详细信息的显示。DataList 控件的属性代码如下:

```
<asp:DataList ID="DataList1" runat="server"
    onitemcommand="DataList1_ItemCommand">
    <ItemTemplate>
        用户名:<%#Eval("stu_name") %>
        <asp:Button ID="Button1" runat="server" Text="详细信息" CommandName=
        "select" />
    </ItemTemplate>
    <SelectedItemTemplate>
        <table border="1">
            <tr>
                <td>学号:</td>
                <td><%#Eval("stu_id") %></td>
            </tr>
            <tr>
                <td>姓名:</td>
                <td><%#Eval("stu_name") %></td>
            </tr>
            <tr>
                <td>性别:</td>
                <td><%#Eval("stu_sex") %></td>
            </tr>
            <tr>
                <td>年龄:</td>
                <td><%#Eval("stu_age") %></td>
            </tr>
            <tr>
                <td colspan="2">
                <asp:Button ID="Button2" runat="server" Text="返回" CommandName=
                "back" /></td>
            </tr>
        </table>
    </SelectedItemTemplate>
</asp:DataList>
```

(3) 当用户单击模板中的按钮时,会引发 DataList 控件的 ItemCommand 事件,在该事件的程序代码中根据不同按钮的 CommandName 属性设置 DataList 控件的 SelectedIndex 属性的值,决定显示详细信息或者取消显示详细信息。最后,重新将控件绑定到数据源,代码如下:

```
protected void DataList1_ItemCommand(object source, DataListCommandEventArgs e)
{   if (e.CommandName =="select")
    {   DataList1.SelectedIndex =e.Item.ItemIndex;
```

```
            Bind();
        }
        if (e.CommandName =="back")
        {
            DataList1.SelectedIndex =-1;
            Bind();
        }
}
```

(4) Bind 函数跟 Page_load 函数代码如下：

```
protected void Page_Load(object sender, EventArgs e)
{   if (!IsPostBack)
    {   Bind();
    }
}
protected void Bind()
{   string strConn =@"Data Source= .\SQLEXPRESS;Initial Catalog=Student;
    Integrated Security=True";
    SqlConnection con =new SqlConnection(strConn);
    con.Open();
    string strSql ="select * from T_students";
    SqlCommand comm =new SqlCommand(strSql, con);
    SqlDataReader dr;
    dr =comm.ExecuteReader();
    DataList1.DataSource =dr;
    DataList1.DataBind();
    dr.Close();
    con.Close();
}
```

运行实例，单击学生姓名则显示学生的详细信息，效果如图 7.30 和图 7.31 所示。

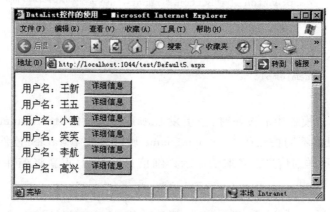

图 7.30　DataList 控件中数据的详细信息显示（一）

图 7.31　DataList 控件中数据的详细信息显示（二）

7.4.6　在 DataList 控件中对数据进行编辑操作

DataList 控件中也可以像 GridView 控件一样，为特定项进行编辑操作。在 DataList 控件中是使用 EditItemTemplate 模板实现这一功能的。下面示例介绍了如何使用 EditItemTemplate 模板对 DataList 控件中的数据项进行编辑。

程序实现的主要步骤如下。

（1）新建一个网站，默认主页为 Default.aspx。在 Default.aspx 页中添加 1 个 DataList 控件。

（2）打开 DataList 控件的项模板编辑模式。在 ItemTemplate 模板中添加 1 个 Label 控件和 1 个 Button 控件；在 EditItemTemplate 模板中添加 2 个 Button 控件、1 个 Label 控件和 3 个 TextBox 控件，代码如下：

```
<asp:DataList ID="DataList1" runat="server"
    oncancelcommand="DataList1_CancelCommand"
    oneditcommand="DataList1_EditCommand"
    onupdatecommand="DataList1_UpdateCommand" OnDeleteCommand=
    "DataList1_DeleteCommand">
<ItemTemplate>
    姓名：<%#Eval("stu_name") %>
    <asp:Button ID="Button1" runat="server" Text="编辑" CommandName="edit" />
</ItemTemplate>
<EditItemTemplate>
    <table><tr><td>学号</td>
        <td><asp:TextBox ID="tid" runat="server" Text='<%#Eval("stu_id") %>'>
            </asp:TextBox></td></tr>
        <tr><td>姓名</td>
            <td><asp:TextBox ID="tname" runat="server" Text='<%#Eval
            ("stu_name") %>'></asp:TextBox></td>
        </tr>
```

```
            <tr><td>性别</td>
                <td><asp:TextBox ID="tpwd" runat="server" Text='<%#Eval
                    ("stu_sex") %>'></asp:TextBox></td>
            </tr>
            <tr><td>年龄</td>
                <td><asp:TextBox ID="TextBox1" runat="server" Text='<%#Eval
                    ("stu_age") %>'></asp:TextBox></td>
            </tr>
            <tr><td><asp:Button ID="Button2" runat="server" Text="更新"
                    CommandName="update" /></td>
                <td><asp:Button ID="Button3" runat="server" Text="取消"
                        CommandName="cancel" />
                        <asp:Button ID="Button4" runat="server" Text="删除"
                    CommandName="delete"/>
                </td>
            </tr>
        </table>
    </EditItemTemplate>
</asp:DataList>
```

注意:"编辑"按钮控件的 CommandName 属性一定要设置为 edit 才能引发 EditCommand 事件,同样"更新"按钮和"取消"按钮的 CommandName 属性也要分别设置为 update、cancel 才能引发 UpdateCommand 事件和 CancelCommand 事件。

(3) 主要程序代码。

当用户单击"编辑"按钮时,将触发 DataList 控件的 EditCommand 事件。在该事件的处理程序中,将用户的选中的项设置为编辑模式,代码如下:

```
protected void DataList1_EditCommand(object source, DataListCommandEventArgs e)
{   //设置 DataList1 控件的编辑项的索引为选择的当前索引值
    DataList1.EditItemIndex =e.Item.ItemIndex;
    Bind();     //重新绑定 DataList 控件
}
```

在编辑模式下,当用户单击"更改"按钮时,将触发 DataList 控件的 UpdataCommand 事件。在该事件的处理程序中,将用户的更改更新到数据库,并取消编辑状态,代码如下:

```
protected void DataList1_UpdateCommand(object source, DataListCommandEventArgs e)
{   string i =DataList1.DataKeys[e.Item.ItemIndex].ToString();
    string name, age,sex;
    name =((TextBox)e.Item.FindControl("tname")).Text;
    age =((TextBox)e.Item.FindControl("tAge")).Text;
    sex =((TextBox)e.Item.FindControl("tSex")).Text;
    string strSql ="update T_students set stu_name='" +name +"',stu_age='" +age +"',
    stu_sex='" +sex +"' where stu_id='" +i +"'";
    string strConn =@"Data Source=.\SQLEXPRESS;Initial Catalog=Student;
    Integrated Security=True";
```

```
        SqlConnection con =new SqlConnection(strConn);
        con.Open();
        SqlCommand comm =new SqlCommand(strSql, con);
        comm.ExecuteNonQuery();
        con.Close();
        DataList1.EditItemIndex =-1;
        Bind();
    }
```

当用户单击"取消"按钮时,将触发 DataList 控件的 CancelCommand 事件。在该事件的处理程序中,取消处于编辑状态的项,并重新绑定数据,代码如下:

```
    protected void DataList1_CancelCommand(object source, DataListCommandEventArgs e)
    {   DataList1.EditItemIndex =-1;       //取消编辑
        Bind();
    }
```

在页面 Page_Load 事件中调用自定义的方法 Bind(),Bind()方法用来绑定 DataList 控件的数据源,代码如下:

```
    protected void Bind()
    {   string strConn =@"Data Source=.\SQLEXPRESS;Initial Catalog=Student;Integrated
        Security=True";
        SqlConnection con =new SqlConnection(strConn);
        con.Open();
        string strSql ="select * from T_students";
        SqlCommand comm =new SqlCommand(strSql, con);    //创建 command 对象
        SqlDataReader dr;
        dr =comm.ExecuteReader();   //command 对象执行相应的 SQL 语句,并把执行的结果存在
                                    //dr 对象中
        DataList1.DataSource =dr;   //设置 DataList 控件的数据源
        DataList1.DataKeyField ="stu_id";
        DataList1.DataBind();       //实现数据绑定
        dr.Close();
        con.Close();
    }
```

注意:DataList 控件在绑定数据时,应先将 DataKeyField 属性设置为数据表的主键。在程序中,可以由 DataKeys 集合利用索引值取得各数据的索引值。

7.4.7 获取 DataList 控件中控件数据的方法

在使用 DataList 控件开发网站时,经常需要获取 DataList 控件中的数据,开发人员可以使用 FindControl()方法来获取 DataList 控件中的数据。例如使用 DataList 控件显示商品信息,当想要修改某一个商品数量时,首先选择要修改的商品,将该商品的数量修改,修改后使用 FindControl()方法获取修改后的数量存储到数据库中。使用 FindControl()方法获取商品数量的主要代码如下:

```
String Name=((TextBox)e.Item.FindControl("TextBoxName")).Text;
```

7.4.8 在 DataList 控件中创建多个列

在 DataList 控件中创建多个列,只要将 DataList 控件的 RepeatColumns 属性设置为需要显示的列数就可以了。例如,将同一字段显示为 2 列,把 DataList 控件的 RepeatColumns 属性设置为 2,显示如图 7.32 所示。

注意：DataList 控件的 RepeatColumns 属性是用来设置列的数目,本实例设置为 3。RepeatDriection 属性是用来设置列的方向,这个属性有两个值：一个是 Horizontal,表示水平方向；另一个是 Vertical,表示垂直方向,本实例使用的是 Horizontal。

图 7.32 DataList 控件中创建多个列

7.5 Repeater 控件

Repeater 控件使用列表方式来显示数据,能够让用户定义 Template 模板标记,自动用模板标记的项目,像循环那样重复编排数据源的数据,其基本语法如下：

```
<asp:Repeater ID="Repeater1" runat="server">
    <HeaderTemplate></HeaderTemplate>
    <ItemTemplate></ItemTemplate>
    <AlternatingItemTemplate></AlternatingItemTemplate>
    <FooterTemplate></FooterTemplate>
</asp:Repeater>
```

上述 Repeater 控件使用 Template 模板标记(标记内容可以使用 HTML 标记)编排数据。各种 Template 标记的说明如表 7.2 所示。

表 7.2 Template 标记的说明

Template 模板标记	说　　明
ItemTemplate	定义列表项目,也就是重复显示部分,对于数据表来说是每条记录,此为必需标记
AlternatingItemTemplate	项目交叉使用不同样式的模板,例如,记录轮流使用不同色彩显示,可以定义此标记,奇数项目(以 0 开始)使用此模板显示,偶数项目使用 ItemTemplate 模板

续表

Template 模板标记	说　明
SeparatorTemplate	项目分隔模板,可以定义分隔标记,通常使用 HTML 标记\<br\>或\<hr\>,如果没有定义就不显示
HeaderTemplate	定义列表标题,对数据表来说,就是\<table\>和记录的标题列,如果没有定义就不显示
FooterTemplate	定义列表脚注,对数据表来说,就是结尾标记\</table\>,如果没有定义就不显示

7.5.1 Repeater 控件以表格显示数据表

Repeater 控件的功能如同 Foreach 语句,换句话说,配合 HTML 表格标记和数据源控件,就可以使用表格来显示数据表的记录数据。

Repeater 控件首先显示 HeaderTemplate 后,依数据表的记录数重复显示 ItemTemplate 和 AlternatingItemTemplate 模板中定义的内容,最后显示 FooterTemplate 模板标记的内容。

在\<td\>标记中每个单元格使用\<%# %\>符号标记为表达式,此例是获取和显示指定数据表字段内容,代码如下:

```
<%#Eval("stu_name")%>
```

上述程序代码的参数字符串是数据表的字段名称。Default.aspx 页面具体代码如下:

```
<asp:Repeater ID="Repeater1" runat="server">
    <HeaderTemplate>
        <table>
            <tr bgcolor="#ffcc99">
                <td>学号</td>
                <td>姓名</td>
                <td>年龄</td>
                <td>性别</td>
            </tr>
    </HeaderTemplate>
    <ItemTemplate>
        <tr>
            <td><%#Eval("stu_id") %></td>
            <td><%#Eval("stu_name") %></td>
            <td><%#Eval("stu_age") %></td>
            <td><%#Eval("stu_sex") %></td>
        </tr>
    </ItemTemplate>
    <AlternatingItemTemplate>
        <tr bgcolor="#dadfdd">
            <td><%#Eval("stu_id") %></td>
            <td><%#Eval("stu_name") %></td>
```

```
            <td><%#Eval("stu_age") %></td>
            <td><%#Eval("stu_sex") %></td>
        </tr>
    </AlternatingItemTemplate>
    <FooterTemplate>
        </table>
    </FooterTemplate>
</asp:Repeater>
```

Default.aspx.cs 页面代码如下：

```
protected void Page_Load(object sender, EventArgs e)
{
    string strConn =@"Data Source=.\SQLEXPRESS;Initial Catalog=Student;
    Integrated Security=True";
    SqlConnection con =new SqlConnection(strConn);
    con.Open();
    string strSql ="select stu_id,stu_name,stu_sex,stu_age from T_students";
    SqlCommand comm =new SqlCommand(strSql, con);
    SqlDataReader dr;
    dr =comm.ExecuteReader();
    Repeater1.DataSource =dr;
    Repeater1.DataBind();
    dr.Close();
    con.Close();
}
```

最终的运行结果如图 7.33 所示。

图 7.33　Repeater 控件的运行结果

7.5.2　Repeater 控件分页显示数据表中数据

以上实例中也介绍过分页显示数据的方法，但是那些都是把数据库中的数据都全部查询出来，再计算显示那个部分，然后进行显示，这样做的弊端就是如果数据库中表中的数据非常庞大,分页显示的效率就会很低,导致显示数据过慢,最终会让客户放弃浏览网页。

现在介绍的新的一种分页思想,仅仅到数据库中把需要显示的记录查询出来,这样显示的速度会非常快,不会导致网页的瘫痪。

pageTest.aspx 的代码如下:

```
<asp:Repeater ID="Repeater1" runat="server">
    <HeaderTemplate>
        <table>
            <tr bgcolor="#ffcc99">
                <td>id</td>
                <td>姓名</td>
                <td>性别</td>
                <td>年龄</td>
            </tr>
    </HeaderTemplate>
    <ItemTemplate>
        <tr>
            <td><%#Eval("stu_id") %></td>
            <td><%#Eval("stu_name") %></td>
            <td><%#Eval("stu_sex") %></td>
            <td><%#Eval("stu_age") %></td>
        </tr>
    </ItemTemplate>
    <AlternatingItemTemplate>
        <tr bgcolor="#dadfdd">
            <td><%#Eval("stu_id") %></td>
            <td><%#Eval("stu_name") %></td>
            <td><%#Eval("stu_sex") %></td>
            <td><%#Eval("stu_age") %></td>
        </tr>
    </AlternatingItemTemplate>
    <FooterTemplate>
        </table>
    </FooterTemplate>
</asp:Repeater>
<asp:Label ID="Label1" runat="server" Text=""></asp:Label>
<asp:HyperLink ID="first" runat="server">第一页</asp:HyperLink>
<asp:HyperLink ID="front" runat="server">上一页</asp:HyperLink>
<asp:HyperLink ID="next" runat="server">下一页</asp:HyperLink>
<asp:HyperLink ID="last" runat="server">末一页</asp:HyperLink>
```

pageTest.aspx.cs 的代码如下:

```
protected void Page_Load(object sender, EventArgs e)
{   int cpage;
```

```csharp
if (Request.QueryString["page"] !=null)
    cpage =Convert.ToInt32(Request.QueryString["page"]);
else
    cpage =1;
int pagesize =2;
string sql ="";
string myConnString = @" Data Source = . \ SQLEXPRESS; Initial Catalog = Student;
Integrated Security=True";
SqlConnection con =new SqlConnection(myConnString);
con.Open();
string strSql ="select count( * ) from T_students";
SqlCommand com =new SqlCommand(strSql, con);
int totalput =Convert.ToInt32(com.ExecuteScalar());
int maxpage =1;
if (totalput %pagesize ==0)
{   maxpage =totalput / pagesize;
}
else
{   maxpage =totalput / pagesize +1;
}
if (maxpage ==0) { maxpage =1; }
if (cpage <1)
{ cpage =1; }
else if (cpage >maxpage)
{ cpage =maxpage; }
if (totalput !=0)
{
  if (cpage ==1)
  {   sql ="select top " +pagesize +" * from T_students order by stu_id desc";
  }
  else
  {   sql ="select top " +pagesize +" * from T_students where stu_id not
      in(select top " + (cpage -1) * pagesize +" stu_id from T_students order
      by stu_id desc) order by stu_id desc";
  }
}
com.CommandText =sql;
SqlDataReader dr;
dr =com.ExecuteReader();
this.Label1.Text ="共有信息" +totalput.ToString() +"条 当前是第" +
cpage.ToString() +"/" +maxpage.ToString() +"页 ";
if (cpage !=1)
{
    this.first.NavigateUrl ="pageTest.aspx?page=1";
    this.front.NavigateUrl ="pageTest.aspx?page=" +Convert.ToString(cpage -1);
```

```
    }
    if (cpage !=maxpage)
    {
        this.next.NavigateUrl ="pageTest.aspx?page=" +Convert.ToString(cpage +1);
        this.last.NavigateUrl ="pageTest.aspx?page=" +maxpage.ToString();
    }
    this.Repeater1.DataSource =dr;
    this.Repeater1.DataBind();
}
```

7.6 小结

本章主要介绍数据控件的使用,第6章讲述的处理数据库中的数据,处理完毕后如何显示在页面当中,就是用到的本章讲述的数据控件。使用数据控件可以把想显示的内容显示处理,同时可以进行数据库的操纵。

课后思考问题

1. GridView 控件、DataList 控件和 Repeater 控件三者之间的区别是什么?从网站的执行效率方面考虑,这3个控件如何能更好地使用?

2. 除了课本上介绍的查询显示方式外,分页显示查询信息有没有更高效的分页显示方式?

第 8 章 单表新闻发布系统的实现

本章要点
- 需求分析。
- 功能描述。
- 数据库说明。
- 项目界面展示。
- 相应的主要代码。

8.1 需求分析

企业网站、门户网站和个人网站等,多数地方需要显示新闻内容,本章介绍的单表新闻系统,就是想让读者首先接受简单的网站的建设流程,使读者很容易地了解网站的制作过程和包含的内容等。

单表新闻发布系统的主要功能为新闻信息的发布,以及新闻信息的浏览。另外参考其他的新闻发布系统,可以将系统分为两部分:一个为后台管理部分,一个为前台显示部分。通过后台管理部分来进行新闻数据的维护,通过前台显示部分进行新闻的浏览。

本系统的主要功能图如图 8.1 所示。

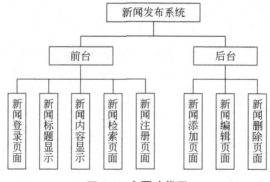

图 8.1 主要功能图

8.2 新闻标题显示

首先介绍首页中新闻标题的显示,显示的结果如图 8.2 所示。

想要显示新闻标题,就需要到数据库中把 News 表中的标题都查找出来,本系统使用三层架构的思想,把对于数据库的操纵都写在类中,首先新建一个类 NewsDB,这个类中包含了对于数据库所有的操作。下面这段代码是对首页中要显示的新闻标题的操作。

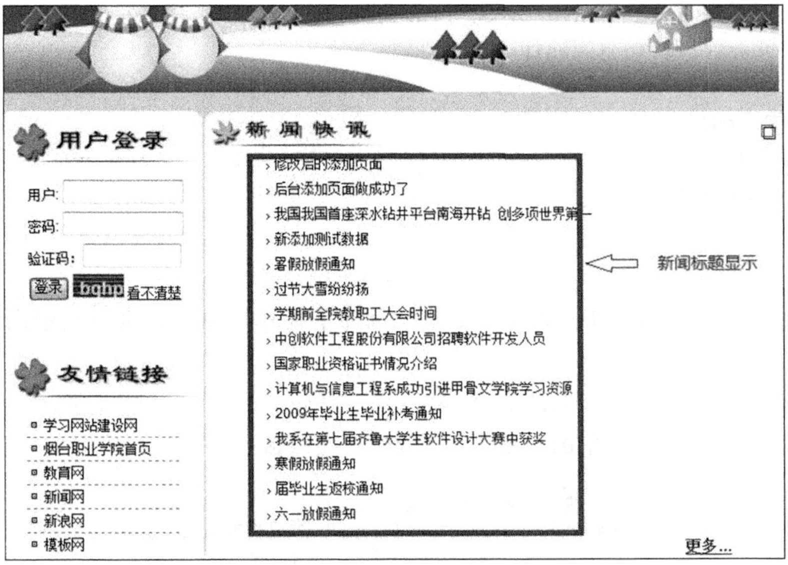

图 8.2 首页新闻标题显示

```
public class NewsDB
{
/// <summary>
/// 显示新闻标题
/// </summary>
/// <returns>新闻标题列表</returns>
public static IList<News>GetList()
{
    //获取连接对象的连接字符串
    string strConn = ConfigurationManager. ConnectionStrings [ " strConn2 "].ConnectionString;
    //建立连接对象
    SqlConnection conn =new SqlConnection(strConn);
    //打开跟数据库的连接
    conn.Open();
    //到数据库中检索前 15 行记录
    string strSql ="select top 15 id,title from [T_news] order by id desc";
    SqlCommand comm =new SqlCommand(strSql, conn);
    //使用 Command 对象执行相应 SQL 语句
    SqlDataReader dr =comm.ExecuteReader();
    //创建一个泛型
    IList<News>newsTotal =new List<News>();
    //把所有的查询结果导到泛型中
    while (dr.Read())
    {
        News news1 =new News();
        news1.ID =Convert.ToInt64(dr["id"]);
```

```
            news1.Title =Convert.ToString(dr["title"]);
            newsTotal.Add(news1);
        }
        dr.Close();
        conn.Close();
        return newsTotal;
        }
}
```

对数据库的操作处理完后,就需要把对数据库的操作结果在首页中显示出来,显示的步骤如下。

(1)在首页中拖动一个 ObjectDataSource 控件,下面配置这个控件的属性,单击 ObjectDataSource 控件右上角的小箭头,进行配置数据源,如图 8.3 所示。

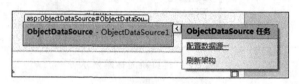

图 8.3　配置数据源

(2)出现配置数据源窗口后,在选择业务对象中选择 NewsDB,也就是上面建立的类 NewsDB。单击"下一步"按钮,如图 8.4 所示。

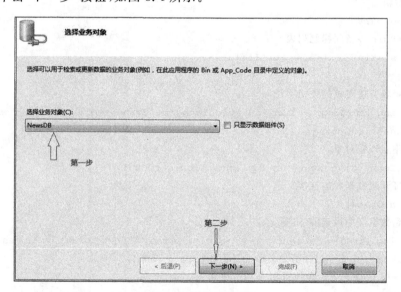

图 8.4　选择业务对象

(3)在出现的定义数据方法窗口中,选择方法 GetList(),返回 IList<News>。最终单击"完成"按钮即可,如图 8.5 所示。

首页中的数据源设置好后,就需要把结果通过控件显示在页面上。这个控件可以使用 Repeater 控件,也可以使用 DataList 控件,也可使用别的相应的数据控件,这里使用 Repeater 控件,对结果进行显示。

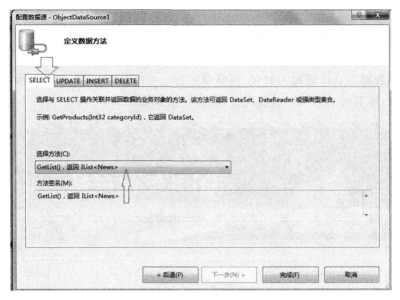

图 8.5 定义数据方法

(1) 在首页相应位置拖动一个 Repeater 控件,单击 Repeater 右上角的箭头,进行数据源的选择,选择好数据源后,就完成了新闻标题的显示,如图 8.6 所示。

图 8.6 Repeater 控件的数据源设置

(2) Repeater 控件中的重要代码如下所示。

```
<div id="main">
    <div id="tips">
        <div id="news">
            <asp:Repeater ID="Repeater1" runat="server" DataSourceID=
            "ObjectDataSource1">
                <ItemTemplate>
                    <ul>
                        <li><a href='<%#Eval("id", "NewsView.aspx?NewsID={0}") %
                            >'><%#Eval("title") %></a></li>
                    </ul>
                </ItemTemplate>
            </asp:Repeater>
            <asp:ObjectDataSource ID="ObjectDataSource1" runat="server"
                SelectMethod="GetList" TypeName="NewsDB"></asp:ObjectDataSource>
        </div>
        <h4><span><a href="NewsMore.aspx">更多…</a></span></h4>
</div>
```

8.3 新闻具体内容的显示

当人们想看哪条具体新闻的时候,需要单击首页中的新闻标题,然后把本新闻的主要的内容显示出来,本节讲述这部分的实现过程,运行结果如图 8.7 所示。

图 8.7 新闻具体内容显示

首先来看首页 Default.aspx 页面中 Repeater 控件中的代码<li><a href='<%＃Eval("id","NewsView.aspx?NewsID={0}")%>'><%＃Eval("title")%></a></li>,当单击某个新闻标题的时候,会转到 NewsView.aspx 这个页面中,转去这个页面后地址后面连接了一个 NewsID 这么一个属性值,人们可以通过取得它的值,从而知道用户到底想看哪一条新闻。

看相应新闻也是需要到数据库中去查询,只要是对数据库的操作就写到 NewsDB 中。重要代码如下:

```
/// <summary>
/// 根据 ID 查询表中的新闻
/// </summary>
/// <param name="id">新闻 ID</param>
/// <returns>新闻对象</returns>
public static News GetNewsByID(long id)
{
    string strConn = ConfigurationManager.ConnectionStrings["strConn2"].ConnectionString;
    SqlConnection conn =new SqlConnection(strConn);
    conn.Open();

    string strSql ="select * from [T_news] where id=@id";
    SqlCommand comm =new SqlCommand(strSql, conn);
    comm.Parameters.AddWithValue("@id",id);
```

```
        SqlDataReader dr = comm.ExecuteReader();

        News new1 = new News();
        while (dr.Read())
        {
            new1.ID = Convert.ToInt64(dr["id"]);
            new1.Title = dr["title"].ToString();
            new1.NewsContent = dr["newsContent"].ToString();
            new1.NewsDate = Convert.ToDateTime(dr["newsDate"]);
            new1.NewsSource = dr["newsSource"].ToString();
        }
        dr.Close();
        conn.Close();
        return new1;
}
```

显示新闻内容的页面使用了母版,显示的格式跟首页相似,其中主要代码如下:

```
<table style="width:90%;margin:auto;">
    <tr><td align="right"><a href="Default.aspx">返回主页</a></td></tr>
    <tr><td align="center">
        <asp:Label ID="newsTitle" runat="server" Text=""></asp:Label>
    </td></tr>
    <tr><td>
        <hr style="height:1px;" />
    </td></tr>
    <tr><td align="right">
        发布时间:<asp:Label ID="newsDate" runat="server" Text=""></asp:Label>
    </td></tr>
    <tr><td>
        <asp:Label ID="newsContent" runat="server" Text=""></asp:Label>
    </td></tr>
    <tr><td align="right">
        新闻来源:<asp:Label ID="newsSource" runat="server" Text=""></asp:Label>
    </td></tr>
</table>
```

新闻显示的.cs代码如下:

```
protected void Page_Load(object sender, EventArgs e)
{
    long id = Convert.ToInt64(Request.QueryString["NewsID"]);
    News newsById = new News();              //接受NewsDB返回的结果
    if (Request.QueryString["NewsID"] == null)
    {
        Response.Write("<script>alert('没有传递ID值!')</script>");
    }
```

```
        else
        {
            newsById = NewsDB.GetNewsByID(id);
            newsTitle.Text = newsById.Title.ToString();
            newsDate.Text = newsById.NewsDate.ToShortTimeString();
            newsContent.Text = newsById.NewsContent.ToString();
            newsSource.Text = newsById.NewsSource.ToString();
        }
    }
```

8.4 新闻检索功能

新闻检索就是到数据库中把用户填写的相关新闻都查找出来,运行的页面如图 8.8 所示。

图 8.8 首页新闻检索

新闻检索结果显示如图 8.9 所示。

图 8.9 新闻检索结果显示

新闻检索对于数据库的操作代码如下:

```
/// <summary>
/// 根据新闻标题检索数据库中的相应数据
/// </summary>
/// <param name="selectTile">查询的标题内容</param>
/// <returns>新闻列表</returns>
public static IList<News> SelectNewsTitle(string selectTile)
{
```

```csharp
string strConn = ConfigurationManager.ConnectionStrings["strConn2"].ConnectionString;
SqlConnection conn = new SqlConnection(strConn);
conn.Open();

string strSql = "select id,title from [T_News] where title like '%" + selectTile + "%'";
//string strSql = "select id,title from [T_News] where title like '%@title%'";
SqlCommand comm = new SqlCommand(strSql, conn);
//comm.Parameters.AddWithValue("@title", selectTile);
SqlDataReader dr = comm.ExecuteReader();

IList<News> SelectNews = new List<News>();
while (dr.Read())
{
    News news1 = new News();
    news1.ID = Convert.ToInt64(dr["id"]);
    news1.Title = dr["title"].ToString();
    SelectNews.Add(news1);
}
dr.Close();
conn.Close();

return SelectNews;
}
```

对于新闻检索结果的显示,方法如同 8.2 章节,在这不再详细介绍。

8.5 新闻后台登录页面实现

前面介绍了部分重要的前台页面的实现,本节介绍后台的登录页面的实现,实现的界面如图 8.10 所示。

图 8.10 登录页面

注册页面主要代码如下：

```csharp
protected void Page_Load(object sender, EventArgs e)
{
    if (!IsPostBack)
    {
        if (Session["UserName"] !=null)
        {
            Session.Abandon();
        }
    }
    if (IsPostBack)
    {
        if (Session["CheckCode"].ToString() ==Request.Form["valid"].ToString())
        {
            Login();
        }
        else
        {
            this.ClientScript.RegisterStartupScript(this.GetType(), "aa",
            "<script>alert('验证码不对,请重新填写!');</script>");
        }

    }
}
protected void Login()
{
    string username =Request.Form["username"].ToString();
    string password =Request.Form["pwd"].ToString();

    if (username =="" || password =="")
    {
        this.ClientScript.RegisterStartupScript(this.GetType(), "aa",
        "<script>alert('用户名或密码不能为空!');</script>");

    }
    else
    {
        if (username =="admin" && password =="123")
        {
            Session["UserName"] ="admin";
            Response.Redirect("AdminIndex.aspx");
        }
        else
        {
            this.ClientScript.RegisterStartupScript(this.GetType(), "aa",
```

```
            "<script>alert('用户名或密码不正确!');</script>");
        }
    }
}
```

8.6 添加新闻

添加新闻是后台页面中必需的部分,动态页面的显示就是需要在后台可以对前台显示的内容进行操作。添加页面的运行结果如图 8.11 所示。

图 8.11 后台添加页面

后台添加页面也是需要对数据库中的表进行添加,所以操作数据库的代码还是写在 NewsDB 中,主要代码如下:

```
/// <summary>
/// 添加新闻
/// </summary>
/// <param name="ne">表单上填写的新闻内容</param>
/// <returns>是否添加成功</returns>
public static bool AddNews(News ne)
{
    string strConn = ConfigurationManager.ConnectionStrings [ " strConn2 "].ConnectionString;
    SqlConnection conn =new SqlConnection(strConn);
    conn.Open();

    string strSql ="insert into [T_News] (title,newsContent,newsSource) values (@title,@content,@source)";
```

```
        SqlCommand comm = new SqlCommand(strSql, conn);
        comm.Parameters.AddWithValue("@title", ne.Title);
        comm.Parameters.AddWithValue("@content", ne.NewsContent);
        comm.Parameters.AddWithValue("@source", ne.NewsSource);

        int i = comm.ExecuteNonQuery();

        conn.Close();

        if (i > 0)
        {
            return True;
        }
        else
        {
            return False;
        }
    }
```

添加页面的主要代码如下:

```
<table>
    <tr>
        <td>新闻标题: </td>
        <td>
            <asp:TextBox ID="txtTitle" runat="server"></asp:TextBox></td>
        <td></td>
    </tr>
    <tr>
        <td>新闻内容: </td>
        <td>
            <asp:TextBox ID="txtContent" runat="server" TextMode="MultiLine"
            Height="361px" Width="497px"></asp:TextBox></td>
        <td></td>
    </tr>
    <tr>
        <td>新闻来源: </td>
        <td>
            <asp:TextBox ID="txtSource" runat="server"></asp:TextBox></td>
        <td></td>
    </tr>
    <tr>
        <td>
            <asp:Button ID="Button1" runat="server" Text="添加" onclick=
            "Button1_Click" /></td>
        <td>
```

```
            <asp:Button ID="Button2" runat="server" Text="重置" onclick=
            "Button2_Click" /></td>
        <td></td>
    </tr>
</table>
```

后台添加页面的.cs页面的代码如下：

```
protected void Button1_Click(object sender, EventArgs e)
{
    News n1 =new News();
    n1.Title =txtTitle.Text;
    n1.NewsContent =txtContent.Text;
    n1.NewsSource =txtSource.Text;
    bool flag =NewsDB.AddNews(n1);

    if (flag ==True)
    {
        this.ClientScript.RegisterStartupScript(this.GetType(), "aa",
        "<script>alert('添加成功');</script>");
        txtTitle.Text ="";
        txtContent.Text ="";
        txtSource.Text ="";
        txtTitle.Focus();        //获取焦点
    }
    else
    {
        this.ClientScript.RegisterStartupScript(this.GetType(), "aa",
        "<script>alert('添加不成功!');</script>");

    }
}
protected void Button2_Click(object sender, EventArgs e)
{
    txtTitle.Text ="";
    txtContent.Text ="";
    txtSource.Text ="";
}
```

8.7 编辑新闻

编辑页面就是对数据库中的新闻表进行修改和删除，运行页面如图8.12所示。
编辑页面首先到数据库中把所有的记录查找出来，进行分页显示，然后再加上编辑和删除两个功能键。

图 8.12 编辑页面

对于数据库的查询,主要代码如下:

```
/// <summary>
/// 获取新闻表中的所有记录
/// </summary>
/// <returns>新闻标题列表</returns>
public static IList<News>GetAllList()
{
    string strConn = ConfigurationManager.ConnectionStrings[" strConn2"].ConnectionString;
    SqlConnection conn =new SqlConnection(strConn);
    conn.Open();

    string strSql ="select id,title from [T_news] order by id desc";
    SqlCommand comm =new SqlCommand(strSql, conn);
    SqlDataReader dr =comm.ExecuteReader();

    IList<News>newsAll =new List<News>();

    while (dr.Read())
    {
        News news1 =new News();
        news1.ID =Convert.ToInt64(dr["id"]);
        news1.Title =Convert.ToString(dr["title"]);
        newsAll.Add(news1);
    }

    dr.Close();
```

```
            conn.Close();
            return newsAll;
      }
```

编辑页面使用了 DataList 控件来实现的,主要代码如下:

```
<asp:DataList ID="DataList1" runat="server" DataSourceID="ObjectDataSource1"
    Width="90%" DataKeyField="id" ondeletecommand="DataList1_DeleteCommand">
    <HeaderTemplate>
        <table class="smallLine" width="100%">
            <tr>
                <td class="smallLine">ID 号</td>
                <td class="smallLine">新闻标题</td>
                <td class="smallLine">编辑</td>
            </tr>
    </HeaderTemplate>
    <ItemTemplate>
        <tr>
            <td class="smallLine"><%#Eval("id") %></td>
            <td class="smallLine"><%#Eval("title")%></td>
            <td class="smallLine">
                <a href='<%#Eval("id","edit.aspx?newsID={0}") %>'>编辑</a>|
                <asp:LinkButton ID="lbDelete" runat="server" CommandName="delete"
                    OnClientClick="return confirm('确定要删除此新闻吗?')">删除
                </asp:LinkButton>
            </td>
        </tr>
    </ItemTemplate>
    <FooterTemplate>
        </table>
    </FooterTemplate>
</asp:DataList>
<asp:ObjectDataSource ID="ObjectDataSource1" runat="server"
    SelectMethod="GetAllList" TypeName="NewsDB">
    </asp:ObjectDataSource>
```

当单击编辑超链接时,转到 edit.aspx 页面,此页面的运行结果如图 8.13 所示。

修改页面需要对数据库中的信息进行修改,所以也得对数据库操作,操作的代码写在 NewsDB 类中,主要的代码如下:

```
/// <summary>
/// 编辑新闻内容
/// </summary>
/// <param name="ne">编辑好的新闻记录</param>
/// <returns>是否编辑成功</returns>
public static bool EditNews(News ne)
{
```

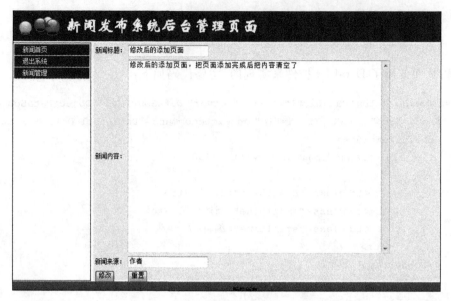

图 8.13 修改页面

```
string strConn = ConfigurationManager.ConnectionStrings[" strConn2 "].ConnectionString;
SqlConnection conn =new SqlConnection(strConn);
conn.Open();

string strSql ="update [T_News] set title=@title,newsContent=@content, newsSource=@source where id=@id";
SqlCommand comm =new SqlCommand(strSql, conn);
comm.Parameters.AddWithValue("@title", ne.Title);
comm.Parameters.AddWithValue("@content", ne.NewsContent);
comm.Parameters.AddWithValue("@source", ne.NewsSource);
comm.Parameters.AddWithValue("@id", ne.ID);

int i =comm.ExecuteNonQuery();

conn.Close();

if (i >0)
{
    return True;
}
else
{
    return False;
}
}
```

编辑页面的主要代码如下:

```
<table>
    <tr>
        <td>新闻标题:</td>
        <td>
            <asp:TextBox ID="txtTitle" runat="server"></asp:TextBox></td>
        <td></td>
    </tr>
    <tr>
        <td>新闻内容:</td>
        <td>
            <asp:TextBox ID="txtContent" runat="server" TextMode="MultiLine"
            Height="361px" Width="497px"></asp:TextBox></td>
        <td></td>
    </tr>
    <tr>
        <td>新闻来源:</td>
        <td>
            <asp:TextBox ID="txtSource" runat="server"></asp:TextBox></td>
        <td></td>
    </tr>
    <tr>
        <td>
            <asp:Button ID="Button1" runat="server" Text="修改" onclick=
            "Button1_Click" /></td>
        <td>
            <asp:Button ID="Button2" runat="server" Text="重置" onclick=
            "Button2_Click" /></td>
        <td></td>
    </tr>
</table>
```

编辑页面的.cs代码如下:

```
protected void Page_Load(object sender, EventArgs e)
{
    if (!IsPostBack)         //第一次加载时读取数据库中的数据
    {
        long id =Convert.ToInt64(Request.QueryString["newsID"]);
        News new1 =new News();
        new1 =NewsDB.GetNewsByID(id);
        if (new1 ==null)
        {
            this.ClientScript.RegisterStartupScript(this.GetType(), "aa",
            "<script>alert('没有查询到相应记录!');</script>");
```

```csharp
            }
            else
            {
                txtTitle.Text = new1.Title;
                txtContent.Text = new1.NewsContent;
                txtSource.Text = new1.NewsSource;
            }
        }
    }

    protected void Button1_Click(object sender, EventArgs e)
    {
        News news1 = new News();
        news1.ID = Convert.ToInt64(Request.QueryString["newsID"]);
        news1.Title = txtTitle.Text;
        news1.NewsContent = txtContent.Text;
        news1.NewsSource = txtSource.Text;

        bool flag = NewsDB.EditNews(news1);

        if (flag == True)
        {
            this.ClientScript.RegisterStartupScript(this.GetType(), "aa",
            "<script>alert('修改成功');location='EditNews.aspx'</script>");
        }
        else
        {
            this.ClientScript.RegisterStartupScript(this.GetType(), "aa",
            "<script>alert('修改不成功!');</script>");

        }
    }
    protected void Button2_Click(object sender, EventArgs e)
    {
        txtTitle.Text = "";
        txtContent.Text = "";
        txtSource.Text = "";
    }
```

最后只剩下删除数据库中的信息。删除时为了避免误删，首先当用户单击"删除"时，要求询问用户是否真的要删除此记录，是的话再进行删除操作，主要代码如下：

```
<asp:LinkButton ID="lbDelete" runat="server" CommandName="delete"
    OnClientClick="return confirm('确定要删除此新闻吗?')">删除
</asp:LinkButton>
```

上面的代码实现的是在删除前先弹出对话框进行确认。对数据库操作的代码如下：

```csharp
/// <summary>
/// 删除相应记录
/// </summary>
/// <param name="id">按照 id 查询要删除的数据</param>
/// <returns>是否删除成功</returns>
public static bool DeleteNews(long id)
{
    string strConn = ConfigurationManager.ConnectionStrings["strConn2"].
    ConnectionString;
    SqlConnection conn =new SqlConnection(strConn);
    conn.Open();

    string strSql ="delete from [T_News] where id=@id";
    SqlCommand comm =new SqlCommand(strSql, conn);
    comm.Parameters.AddWithValue("@id", id);

    int i =comm.ExecuteNonQuery();

    conn.Close();

    if (i >0)
    {
        return True;
    }
    else
    {
        return False;
    }
}
```

编辑页面中进行删除处理的代码如下：

```csharp
protected void DataList1_DeleteCommand(object source, DataListCommandEventArgs e)
{
    long i =Convert.ToInt64(DataList1.DataKeys[e.Item.ItemIndex]);

    bool flag =NewsDB.DeleteNews(i);

    if (flag ==True)
    {
        DataList1.DataSourceID ="ObjectDataSource1";
        DataList1.DataBind();
    }
```

```
        else
        {
            this.ClientScript.RegisterStartupScript(this.GetType(), "aa",
            "<script>alert('删除不成功!');</script>");

        }
    }
```

8.8 小结

本章主要讲述一个真实的项目的制作过程。讲述了单个表的新闻发布系统的实现方式。分别从前台的制作，到后台的制作进行了详细的讲述。

第 9 章　新闻发布系统

本章要点
- 需求分析。
- 功能描述。
- 数据库说明。
- 项目界面展示。
- 相应的主要代码。

9.1　需求分析

现代新闻学诞生有 200 多年的历史了,自从造纸术和印刷术的出现,新闻学的发展脚步就没有一刻停歇下来,随着技术的不断进步,新闻也在不断地发生变化,从早期的纸张记录,到蒸汽印刷机带来的报业繁荣,乃至新闻电讯稿在美国内战期被广泛使用,随着收音机的兴起,人们听到了更多梦寐以求的声音,电视台、卫星电视的出现,改变了人们的生活,到了今天的网络时代,人们甚至只需一台计算机和一根电话线就可以看到世界任何一处的信息。在不久的未来,相信手机将为新闻带来新的纪元。

网站新闻发布系统又称为信息发布系统,是将网页上的某些需要经常变动的信息,类似新闻、新产品发布和业界动态等更新信息集中管理,并通过信息的某些共性进行分类,最后系统化、标准化发布到网站上的一种网站应用程序。网站信息通过一个操作简单的界面加入数据库,然后通过已有的网页模板格式与审核流程发布到网站上。它的出现大大减轻了网站更新维护的工作量,通过网络数据库的引用,将网站的更新维护工作简化到只需录入文字和上传图片,从而使网站的更新速度大大缩短,在某些专门的网上新闻站点,如新浪的新闻中心等,新闻的更新速度已经是即时更新,从而大大加快了信息的传播速度,也吸引了更多的长期用户群,时时保持网站的活动力和影响力。

新闻发布系统的主要功能为新闻信息的发布,以及新闻信息的浏览,另外参考其他的新闻发布系统,可以将系统分为两个部分:一个为后台管理部分;另一个为前台显示部分。通过后台管理部分来进行新闻数据的维护,通过前台显示部分进行新闻的浏览。

本系统主要提供图片上传、浏览、管理专辑等相关功能。

1. 需求中涉及的用户类型

(1) 系统维护的员工。
(2) 相关人员。
(3) 前台浏览信息的用户。

2. 不同类型用户所进行的操作

(1) 系统维护的员工。
① 添加、删除相关人员。

② 指定相关人员权限。
③ 添加、删除、修改栏目。
④ 添加、删除、修改类别。
⑤ 调整栏目、类别的顺序。
⑥ 发布新闻。
⑦ 管理日志。
⑧ 分析统计每日、每月的新闻信息。
⑨ 修改密码。
(2) 相关人员。
① 发布新闻信息。
② 分析统计每日、每月的新闻信息。
③ 修改密码。
(3) 企业用户。
浏览新闻。

3. 分析结果

通过分析上面的需求，可以得到相应的角色的功能职责范围，使用系统的人可以分为3种：第一种为后台部分的系统管理员，第二种为后台部分的普通用户，第三种为前台部分的浏览器终端用户。

各用户角色及职责的详细描述如表9.1所示。

表 9.1 角色名称及职责描述

角色名称	职责描述
后台系统管理员用户	维护网站信息，维护新闻分类，维护新闻，维护普通用户基本信息
后台管理系统普通用户	维护新闻信息，查看当日新闻列表，根据用户权限（用户管理的新闻栏目）维护栏目下的新闻，包括新闻信息的添加、修改、删除、查询浏览等功能
前台浏览器终端用户	浏览查看新闻

9.2 功能描述

9.2.1 后台登录

新闻发布系统要实现后台新闻管理、栏目管理、用户管理等一系列功能，而这些功能的实现之前需要实现的是系统的登录，新闻发布系统要求登录后台新闻管理的人员有两种：一种是普通后台用户；另一种是后台系统管理员。后台普通用户可以管理该用户所管辖的新闻分类下的新闻数据，可以添加、修改、删除新闻。并且可以浏览当日所有发布的新闻；后台管理员可以进行除普通后台用户的操作之外，还要能够进行用户管理、栏目管理、类别管理、日志管理、查看网站配置信息以及新闻流量月统计等操作，登录系统时需要区分这两种用户。

9.2.2 新闻栏目和类别管理

新闻类别属于新闻栏目,发布新闻时,选择新闻类别即可。栏目要求能够添加、删除、修改。类别也是同样的要求,另外修改类别时要求能够调整其所属的栏目。新闻栏目以及新闻类别显示顺序要求能够进行修改。此功能由后台管理员进行操作,后台普通用户没有权限操作。

9.2.3 用户管理

新闻发布系统的最终用户划分为三类,专门负责系统维护的人划分为后台管理员,将发布新闻信息的用户划分为后台普通用户,将公司员工及浏览网页者定位为系统的终端用户。

后台管理员只有一个,初始化到数据库中,在系统中此用户不能被修改、删除,但是可以修改密码。后台管理员可以添加后台普通用户,并且可以修改、删除用户,还可以维护这些用户的权限。后台普通用户不能添加后台普通用户。所有后台用户只允许修改自己的密码。

后台管理员可以修改普通用户的权限,权限的形式为某个后台普通用户是否拥有某个新闻类别的管理权。系统不为普通用户设置栏目权限。后台普通用户不能修改权限。

终端用户不需要添加到数据库中,直接进行前台页面的浏览即可。

9.2.4 新闻发布

要发布的新闻的内容包括标题、所属类别、发布时间、发布人、来源、关键字、内容。其中发布时间取服务器当前时间,发布人取当前登录用户。其他新闻项目输入即可。

已发布的新闻能够修改,已发布的新闻以列表形式显示(该列表要求提供查询,能够按照类别以及关键字进行新闻的查询),然后通过列表打开相应的新闻修改页面,保存数据时发布时间改为当前的服务器时间,发布人不再进行修改保持原有数据即可。其他的新闻项目以修改后的数据更新。

9.2.5 日志管理、流量统计及当日新闻查看

为了系统的安全,为了能够了解登录系统的都有那些用户,所以需要进行日志的记录。当用户登录系统时需要记录下来用户的ID、登录时间以及登录IP,并且可以对这些日志进行删除。

为了了解当月最热点的新闻,需要有新闻月流量的统计,可以查看新闻每月点击率排行,排行榜按照从多到少的顺序排列。

发布新闻时如果其他的用户发布了相类似的新闻,而当前登录用户尚不了解最新的新闻,这样很容易造成新闻的重复发布。为了防止此种情况,系统需要展示给当前登录用户今日发布信息的列表,以进行快速地浏览新发布的新闻信息。

9.2.6 前台显示页面

前台页面首页要将所有的栏目都显示出来,并且每个栏目下要显示本栏目下最新的10

条新闻。新闻列表下要有"更多…"的链接,用于打开本栏目的页面。栏目页面包括其下所有的类别,类别下同样显示本类别下最新的 10 条新闻。新闻列表下也要具有"更多…"的链接,用于打开本类别的页面。类别页面包括了此类别下的所有新闻,并且提供分页功能,新闻以时间倒序排序。以上新闻列表均能单击标题进入新闻浏览页面,进行新闻的详细信息的浏览。

9.3 数据库说明

首先新建数据库 News_Manage,然后新建系统用户数据表 t_News_User,该表各字段如表 9.2 所示。

表 9.2 t_News_User(用户信息表)

列　名	描　述	数据类型(精度范围)	空/非空	约束条件
USERID	用户 ID	int	非空	PK(自增)
USERNAME	用户名称	varchar(50)	空	
USERPASSWORD	用户密码	varchar(256)	空	
POWER	是否是管理员	bit	空	

t_Item 表各字段如表 9.3 所示。

表 9.3 t_Item(新闻栏目表)

列　名	描　述	数据类型(精度范围)	空/非空	约束条件
ITEMID	栏目 ID	int	非空	PK(自增)
ITEMNAME	栏目名称	varchar(50)	空	
ITEMDESC	栏目描述	varchar(200)	空	
ITEMORDER	栏目顺序	int	空	

t_Class 表如表 9.4 所示。

表 9.4 t_Class(新闻类别表)

列　名	描　述	数据类型(精度范围)	空/非空	约束条件
CLASSID	类别 ID	int	非空	PK(自增)
CLASSNAME	类别名称	varchar(50)	空	
CLASSDESC	类别描述	varchar(200)	空	
CLASSORDER	类别顺序	int	空	
ITEMID	所属栏目 ID	int	空	

t_Popedom 表如表 9.5 所示。

表 9.5 t_Popedom（用户信息表）

列　名	描　述	数据类型（精度范围）	空/非空	约束条件
POPEDOMID	权限 ID	int	非空	PK（自增）
CLASSID	新闻类别 ID	int	空	FK
USERID	用户 ID	int	空	FK

t_News 表如表 9.6 所示。

表 9.6 t_News（新闻表）

列　名	描　述	数据类型（精度范围）	空/非空	约束条件
NEWSID	权限 ID	int	非空	PK（自增）
NEWSTITLE	新闻类别 ID	varchar(100)	空	
CLASSID	用户 ID	int	空	FK
NEWSDATE	发布时间	datetime	空	
NEWSKEY	关键字	varchar(20)	空	
NEWSOURCE	新闻来源	varchar(100)	空	
NEWSCONTENT	新闻内容	ntext	空	
USERID	发布人 ID	int	空	FK
HITS	总点击率	int	空	
MONTHHITS	月点击率	int	空	

t_Log 表如表 9.7 所示。

表 9.7 t_Log（日志表）

列　名	描　述	数据类型（精度范围）	空/非空	约束条件
LOGID	日志 ID	int	空	FK
LOGINIP	登录 IP	varchar(15)	空	
LOGINDATE	登录时间	datetime	空	
USERID	用户 ID	int	空	

9.4　相应的主要代码

9.4.1　前台代码

本系统使用的是三层架构来实现新闻浏览的功能，前台总共分为主页（也就是首页）、分类页面（也就是对于每个分类有更加详细的显示）、新闻显示页面（就是当单击某条新闻时显示具体新闻的页面)。

首页的主要代码如下所示,因为主页的显示用的是母版页,关于母版页的设计就不一一介绍了,读者可以参考其他书籍。

```
<asp:DataList ID="dtlIndex" runat="server" DataSourceID="odsNewsItem"
    RepeatColumns="3" RepeatDirection="Horizontal" Width="672px"
    CellSpacing="2" ShowFooter="False" ShowHeader="False">
    <EditItemStyle VerticalAlign="Top" Width="221px" />
    <ItemStyle HorizontalAlign="Left" VerticalAlign="Top" Width="220px" />
    <ItemTemplate>
        <table style="border: 1px solid #CCCCCC; width:100%;" border="0"
            cellpadding="0" cellspacing="0">
            <tr>
                <td style="background-color: #EEEEEE; border-bottom-style:
                    solid; border-bottom-width: 1px; border-bottom-color:
                    #CCCCCC;" height="20">
                    <asp:Image ID="Image1" runat="server" ImageUrl="~/Images/
                    z1.jpg" /> 
                    <asp:Label ID="lblItemName" runat="server" style=
                    "font-weight: 700; font-size: small" Text='<%#Eval("ItemName")
                    %>'></asp:Label>
                    <asp:HiddenField ID="hfItemID" runat="server" Value=
                    '<%#Eval("ItemID") %>' />
                </td>
            </tr>
            <tr>
                <td Height="220px" valign="top">
                    <asp:DataList ID="DataList2" runat="server" DataSourceID=
                    "odsNews" ShowFooter="False" ShowHeader="False" Width="220px">
                    <EditItemStyle HorizontalAlign="Left" />
                    <ItemStyle HorizontalAlign="Left" VerticalAlign="Top" />
    <ItemTemplate>
        <asp:HyperLink ID="HyperLink1" runat="server" NavigateUrl='<%#Eval
        ("NewsID", "NewsView.aspx?NewsID={0}") %>' Target="_blank" Text=
        '<%#Eval("NewsTitle") %>'></asp:HyperLink>
        <br />
    </ItemTemplate>
</asp:DataList>
    <asp:ObjectDataSource ID="odsNews"
    runat="server"SelectMethod="GetIndexList"
    TypeName="NewsManage.DAL.NewsDB">
    <SelectParameters>
        <asp:ControlParameter ControlID="hfItemID" DefaultValue="%" Name=
        "itemID" PropertyName="Value" Type="Int32" />
    </SelectParameters>
```

```
        </asp:ObjectDataSource>
            </td>
        </tr>
        <tr>
            <td style="text-align: right" height="20">
                <asp:HyperLink ID="HyperLink2" runat="server" NavigateUrl=
                '<%#Eval("ItemID", "ItemNews.aspx?ItemID={0}") %>' Target=
                "_self">更多</asp:HyperLink>
                <asp:Image ID="Image2" runat="server" ImageUrl="~/Images/z7.jpg" />
            </td>
        </tr>
    </table>
</ItemTemplate>
</asp:DataList>
<asp:ObjectDataSource ID="odsNewsItem" runat="server" SelectMethod="GetAll"
TypeName="NewsManage.DAL.NewsItemDB"></asp:ObjectDataSource>
```

最后主页中使用了 ObjectDataSource 控件，引用的是类 NewsItemDB 下的方法，关于 NewsItemDB 的具体代码如下：

```
using System.Data.SqlClient;
using NewsManage.Model;
namespace NewsManage.DAL
{   /// <summary>
    ///ItemDB 的摘要说明
    /// </summary>
    public class NewsItemDB
    {   /// <summary>
        /// 得到所有栏目
        /// </summary>
        /// <returns>栏目列表</returns>
        public static IList<NewsItem>GetAll()
        {
            IList<NewsItem>newsItemList =new List<NewsItem>();
            using (SqlConnection connection = new SqlConnection (DBConnection.
            ConnectString))
            {
                SqlCommand command = new SqlCommand("select * from t_Item order by
                itemorder,itemid", connection);
                connection.Open();
                using (SqlDataReader itemReader =command.ExecuteReader())
                {
                    while (itemReader.Read())
                    {
                        NewsItem newsItem =new NewsItem();
                        newsItem =FillData(itemReader);
```

```csharp
                    newsItemList.Add(newsItem);
                }
            }
        }
        return newsItemList;
    }

    /// <summary>
    /// 添加栏目
    /// </summary>
    /// <param name="itemName">栏目名称</param>
    /// <param name="itemDescription">栏目描述</param>
    /// <param name="itemOrder">栏目顺序</param>
    /// <returns>是否成功</returns>
    public static bool AddItem(string itemName, string itemDescription, int itemOrder)
    {
        using (SqlConnection connection = new SqlConnection(DBConnection.ConnectString))
        {
            SqlCommand command = new SqlCommand("insert into t_item values (@itemname,@itemdesc,@itemorder)", connection);
            command.Parameters.Add("@itemname", SqlDbType.VarChar, 50);
            command.Parameters.Add("@itemdesc", SqlDbType.VarChar, 200);
            command.Parameters.Add("@itemorder", SqlDbType.Int);
            command.Parameters[0].Value = itemName;
            command.Parameters[1].Value = itemDescription;
            command.Parameters[2].Value = itemOrder;
            connection.Open();

            int addItemRows = command.ExecuteNonQuery();
            if (addItemRows > 0)
            {
                return True;
            }
            else
            {
                return False;
            }
        }
    }

    /// <summary>
    /// 删除栏目
    /// </summary>
```

```csharp
/// <param name="itemID">栏目 ID</param>
public static void DeleteItem(int itemID)
{
    using (SqlConnection connection = new SqlConnection(DBConnection.ConnectString))
    {
        SqlCommand commandSelect = new SqlCommand("select count(*) from t_class where itemid=@itemid", connection);
        SqlCommand commandDelete = new SqlCommand("delete from t_item where itemid=@itemid", connection);
        connection.Open();
        commandSelect.Parameters.Add("@itemid", SqlDbType.Int);
        commandSelect.Parameters[0].Value = itemID;
        commandDelete.Parameters.Add("@itemid", SqlDbType.Int);
        commandDelete.Parameters[0].Value = itemID;
        int classRows = Convert.ToInt32(commandSelect.ExecuteScalar());
        if (classRows > 0)
        {
            HttpContext.Current.Response.Write("<script>alert('该栏目下有类别数据,不能删除!');location.replace(window.location.href);</script>");
        }
        else
        {
            int deleteRows = commandDelete.ExecuteNonQuery();
            if (deleteRows > 0)
            {
                //HttpContext.Current.Response.Write("<script>alert('删除成功!');location.replace(window.location.href);</script>");
            }
            else
            {
                HttpContext.Current.Response.Write("<script>alert('删除失败,请稍后重试!');location.replace(window.location.href);</script>");
            }
        }
    }
}

/// <summary>
/// 更新栏目信息
/// </summary>
/// <param name="itemID">栏目 ID</param>
/// <param name="itemName">栏目名称</param>
```

```csharp
/// <param name="itemDescription">栏目描述</param>
/// <param name="itemOrder">栏目顺序</param>
public static void UpdateItem ( int itemID, string itemName, string
itemDescription, int itemOrder)
{
    using (SqlConnection connection = new SqlConnection (DBConnection.
    ConnectString))
    {
        SqlCommand command = new SqlCommand("update t_item set itemname=
        @itemname, itemdesc=@itemdescription, itemorder=@itemorder where
        itemid=@itemid", connection);
        connection.Open();
        command.Parameters.Add("@itemid", SqlDbType.Int);
        command.Parameters.Add("@itemname", SqlDbType.VarChar, 50);
        command.Parameters.Add("@itemdescription", SqlDbType.VarChar, 200);
        command.Parameters.Add("@itemorder", SqlDbType.Int);
        command.Parameters[0].Value = itemID;
        command.Parameters[1].Value = itemName;
        command.Parameters[2].Value = itemDescription;
        command.Parameters[3].Value = itemOrder;
        int updateRows = command.ExecuteNonQuery();
        if (updateRows > 0)
        {
            //HttpContext.Current.Response.Write("<script>alert('更新成
            功!');location.replace(window.location.href);</script>");
        }
        else
        {
            HttpContext.Current.Response.Write("<script>alert('更新失败,请稍后
            重试!');location.replace(window.location.href);</script>");
        }
    }
}

private static NewsItem FillData(IDataReader itemReader)
{
    NewsItem newsItem = new NewsItem();
    newsItem.ItemID = itemReader.GetInt32(itemReader.GetOrdinal("itemid"));
    newsItem.ItemName = itemReader.GetString(itemReader.GetOrdinal
    ("itemname"));
    newsItem.ItemDescription = itemReader.GetString(itemReader.GetOrdinal
    ("itemdesc"));
    newsItem.ItemOrder = itemReader.GetInt32(itemReader.GetOrdinal
    ("itemorder"));
    return newsItem;
```

```
            }
      }
}
```

上面是首页的主要代码,下面把如何显示具体新闻的代码列出,其他页的内容根据这两个页的提示,读者自己查阅源代码。显示新闻的页面代码如下:

```
<table style="width:100%;">
    <tr>
        <td colspan="3" style="text-align: center">
            <asp:Label ID="lblNewsTitle" runat="server" SkinID="NewsTitle">
            </asp:Label>
        </td>
    </tr>
    <tr>
        <td colspan="3" style="text-align: center">
            <hr size="1" style="height: -13px" />
        </td>
    </tr>
    <tr>
        <td style="text-align: center" width="33%" class="style3">
            <span class="style1">所属栏目:</span><asp:Label ID="lblItemName"
            runat="server" SkinID="NewsInfo" style="font-size: 12px"></asp:Label
            >
        </td>
        <td style="text-align: center" width="33%" class="style3">
            <span class="style1">所属类别:</span><asp:Label ID="lblClassName"
            runat="server" SkinID="NewsInfo"></asp:Label>
        </td>
        <td style="text-align: center" title="34%" class="style3">
            <span class="style1">发布时间:</span><asp:Label ID="lblNewsDate"
            runat="server" SkinID="NewsInfo"></asp:Label>
        </td>
    </tr>
    <tr>
        <td colspan="3" style="text-align: center" width="33%" class="style2">
            <span class="style1">搜索关键字:</span><asp:Label ID="lblKeywords"
            runat="server" SkinID="NewsInfo"></asp:Label>
                         
              <span class="style1">新闻来源:</span><asp:Label ID=
            "lblNewsSource"
            runat="server" SkinID="NewsInfo"></asp:Label>
            <hr size="1" style="height: -13px" />
        </td>
    </tr>
    <tr>
```

```html
            <td colspan="3" style="text-align: left">
                <asp:Label ID="lblContent" runat="server" Width="100%"></asp:Label>
            </td>
        </tr>
        <tr>
            <td colspan="3" style="text-align: right" class="style2">
                <span class="style1">新闻作者:</span><asp:Label ID="lblUserName"
                runat="server" SkinID="NewsInfo"></asp:Label>
            </td>
        </tr>
</table>
```

这一页的.cs文件中的代码如下:

```csharp
using NewsManage.DAL;
using NewsManage.Model;

public partial class NewsView : System.Web.UI.Page
{
    protected void Page_Load(object sender, EventArgs e)
    {
        News news = new News();
        if (Request.QueryString["NewsID"] != null)
        {
            news = NewsDB.GetSingle(Convert.ToInt32(Request.QueryString
            ["NewsID"].ToString()));
            if (NewsDB.UpdateHits(news.NewsID))
            {
                lblNewsTitle.Text = news.NewsTitle;
                lblItemName.Text = news.ItemName;
                lblClassName.Text = news.ClassName;
                lblNewsDate.Text = news.NewsDate.ToString("yyyy年MM月dd日");
                lblKeywords.Text = news.NewsKeywords;
                lblNewsSource.Text = news.NewsSource;
                lblContent.Text = HttpUtility.HtmlDecode(news.NewsContent);
                lblUserName.Text = news.UserName;
                this.Title = news.NewsTitle;
            }
            else
            {
                Page.RegisterStartupScript("err1", "<script>alert('打开新闻失败!
                ');window.opener='anyone';window.close();</script>");
            }
        }
        else
        {
```

```
            Page.RegisterStartupScript("err1", "<script>alert('参数不能为空!');
            window.opener='anyone';window.close();</script>");
        }
    }
}
```

代码的最开头进入了两个命名空间，这两个命名空间是在建立类的时候自己起的命名空间，因为在这个页面中要用到类中的方法，所以要把命名空间首先添加进来。这个页中引用到 NewsDB 这个类，下面把这个类中的定义显示如下：

```
using System.Data.SqlClient;
using NewsManage.Model;
namespace NewsManage.DAL
{   /// <summary>
    /// 添加新闻
    /// </summary>
    /// <param name="newsTitle">新闻标题</param>
    /// <param name="classID">所属类别 ID</param>
    /// <param name="newsDate">发布时间</param>
    /// <param name="newsKeywords">关键字</param>
    /// <param name="newsSource">新闻来源</param>
    /// <param name="newsContent">新闻内容</param>
    /// <param name="userID">发布人 ID</param>
    /// <returns></returns>
    public static bool AddNews(string newsTitle,int classID,DateTime newsDate,
    string newsKeywords,string newsSource,string newsContent,int userID)
    {
        using (SqlConnection connection =new SqlConnection(DBConnection.ConnectString))
        {
            SqlCommand command =new SqlCommand("insert into t_news values
            (@newstitle,@classid,@newsdate,@newskey,@newssource,@newscontent,
            @userid,0,0)", connection);
            command.Parameters.AddWithValue("newstitle", newsTitle);
            command.Parameters.AddWithValue("classid", classID);
            command.Parameters.AddWithValue("newsdate", newsDate);
            command.Parameters.AddWithValue("newskey", newsKeywords);
            command.Parameters.AddWithValue("newssource", newsSource);
            command.Parameters.AddWithValue("newscontent", newsContent);
            command.Parameters.AddWithValue("userid", userID);
            connection.Open();
            int insertRows =command.ExecuteNonQuery();
            if (insertRows >0)
            {
                return True;
            }
            else
```

```csharp
            {
                return False;
            }
        }
    }
    /// <summary>
    /// 删除新闻
    /// </summary>
    /// <param name="newsID">新闻 ID</param>
    public static void DeleteNews(int newsID)
    {
        using (SqlConnection connection =new SqlConnection(DBConnection.ConnectString))
        {
            SqlCommand command =new SqlCommand("delete from t_news where newsid=@newsid", connection);
            command.Parameters.AddWithValue("newsID", newsID);
            connection.Open();
            command.ExecuteNonQuery();
        }
    }
    /// <summary>
    /// 得到新闻列表
    /// </summary>
    /// <param name="classID">类别 ID</param>
    /// <param name="newsKeywords">关键字</param>
    /// <returns>新闻列表</returns>
    public static IList<News>GetList(string classID, string newsKeywords)
    {
        string getNewsList;
        Administrator admin = (Administrator)HttpContext.Current.Session["CurrentUser"];
        if (admin.Power)
        {
            getNewsList ="select a.newsid,a.newstitle,isnull(a.newsource,
            ' ') as newsource,isnull(a.newskey,' ') as newskey,
            a.userid,a.hits,a.monthhits,a.classid,a.newsdate,isnull
            (a.newscontent,'') as newscontent,b.itemid,b.itemname,c.classname,
            d.username from t_news a,t_item b,t_class c,t_news_user d where
            a.classid=c.classid and b.itemid=c.itemid and a.userid=d.userid and
            a.classid like @classid and isnull(a.newskey,'') like @newsKeywords
            order by newsdate desc,newsid desc";
        }
        else
        {
            getNewsList ="select a.newsid,a.newstitle,isnull(a.newsource,
```

```csharp
            ' ') as newsource,isnull(a.newskey,' ') as newskey,
            a.userid,a.hits,a.monthhits,a.classid,a.newsdate,isnull
            (a.newscontent,'') as newscontent,b.itemid,b.itemname,c.classname,
            d.username from t_news a,t_item b,t_class c,t_news_user d where
            a.classid=c.classid and b.itemid=c.itemid and a.userid=d.userid and
            a.classid like @classid and isnull(a.newskey,'') like @newsKeywords
            and a.classid in(select classid from t_popedom where userid="+
            admin.UserID.ToString()+") order by newsdate desc,newsid desc";
    }
    IList<News>newsList =new List<News>();
    using (SqlConnection connection =new SqlConnection(DBConnection.ConnectString))
    {
        connection.Open();
        SqlCommand command =new SqlCommand(getNewsList, connection);
        command.Parameters.AddWithValue("@classid", classID);
        command.Parameters.AddWithValue("@newsKeywords", "%" +newsKeywords +"%");
        using (SqlDataReader newsReader =command.ExecuteReader())
        {
            while (newsReader.Read())
            {
                News news =new News();
                news =FillData(newsReader);
                newsList.Add(news);
            }
        }
    }
    return newsList;
}
/// <summary>
/// 得到当日新闻列表
/// </summary>
/// <returns>新闻列表</returns>
public static IList<News>GetTodayList()
{
    string getNewsList;
    getNewsList ="select a.newsid,a.newstitle,isnull(a.newsource,' ')
    as newsource,isnull(a.newskey,' ') as newskey,a.userid,a.hits,
    a.monthhits,a.classid,a.newsdate,isnull(a.newscontent,'') as newscontent,
    b.itemid,b.itemname,c.classname,d.username from t_news a,t_item b,
    t_class c,t_news_user d where a.classid=c.classid and b.itemid=c.itemid
    and a.userid=d.userid and a.newsdate between CONVERT(varchar, getdate(),
    111) and CONVERT(varchar, getdate()+1,111) order by newsdate desc,newsid desc";
    IList<News>newsList =new List<News>();
    using (SqlConnection connection =new SqlConnection(DBConnection.ConnectString))
    {
```

```csharp
            connection.Open();
            SqlCommand command = new SqlCommand(getNewsList, connection);
            using (SqlDataReader newsReader = command.ExecuteReader())
            {
                while (newsReader.Read())
                {
                    News news = new News();
                    news = FillData(newsReader);
                    newsList.Add(news);
                }
            }
        }
        return newsList;
    }
    /// <summary>
    /// 得到月流量统计列表
    /// </summary>
    /// <returns>新闻列表</returns>
    public static IList<News> GetMonthNews()
    {
        string getNewsList;
        getNewsList = "select a.newsid,case when len(a.newstitle)>15 then left
        (a.newstitle,15)+'…' else a.newstitle end as newstitle,isnull(a.newsource,
        ' ') as newsource,isnull(a.newskey,' ') as newskey,a.userid,
        a.hits,a.monthhits,a.classid,a.newsdate,isnull(a.newscontent,'') as
        newscontent,b.itemid,b.itemname,c.classname,d.username from t_news a,
        t_item b,t_class c,t_news_user d where a.classid=c.classid and b.itemid=
        c.itemid and a.userid=d.userid order by monthhits desc,newsid desc";
        IList<News> newsList = new List<News>();
        using (SqlConnection connection = new SqlConnection(DBConnection.ConnectString))
        {
            connection.Open();
            SqlCommand command = new SqlCommand(getNewsList, connection);
            using (SqlDataReader newsReader = command.ExecuteReader())
            {
                while (newsReader.Read())
                {
                    News news = new News();
                    news = FillData(newsReader);
                    newsList.Add(news);
                }
            }
        }
        return newsList;
    }
```

```csharp
/// <summary>
/// 得到首页新闻列表
/// </summary>
/// <param name="itemID">新闻栏目 ID</param>
/// <returns>新闻列表</returns>
public static IList<News>GetIndexList(int itemID)
{
    string getNewsList;
    getNewsList ="select top 10 a.newsid,case when len(a.newstitle)>15 then
    left(a.newstitle,15)+'…' else a.newstitle end as newstitle,isnull
    (a.newsource,' ') as newsource,isnull(a.newskey,' ') as newskey,
    a.userid,a.hits,a.monthhits,a.classid,a.newsdate,isnull(a.newscontent,'')
    as newscontent,b.itemid,b.itemname,c.classname,d.username from t_news a,
    t_item b,t_class c,t_news_user d where a.classid=c.classid and b.itemid=
    c.itemid and a.userid=d.userid and b.itemid=@itemid order by newsdate
    desc,newsid desc";
    IList<News>newsList =new List<News>();
    using (SqlConnection connection =new SqlConnection(DBConnection.ConnectString))
    {
        connection.Open();
        SqlCommand command =new SqlCommand(getNewsList, connection);
        command.Parameters.AddWithValue("@itemid", itemID);
        using (SqlDataReader newsReader =command.ExecuteReader())
        {
            while (newsReader.Read())
            {
                News news =new News();
                news =FillData(newsReader);
                newsList.Add(news);
            }
        }
    }
    return newsList;
}
/// <summary>
/// 得到栏目页新闻列表
/// </summary>
/// <param name="itemID">新闻栏目 ID</param>
/// <returns>新闻列表</returns>
public static IList<News>GetItemList(int classID)
{
    string getNewsList;
    getNewsList ="select top 10 a.newsid,case when len(a.newstitle)>15 then
    left(a.newstitle,15)+'…' else a.newstitle end as newstitle,isnull
    (a.newsource,' ') as newsource,isnull(a.newskey,' ') as newskey,
```

```csharp
            a.userid,a.hits,a.monthhits,a.classid,a.newsdate,isnull(a.newscontent,'')
            as newscontent,b.itemid,b.itemname,c.classname,d.username from t_news
            a,t_item b,t_class c,t_news_user d where a.classid=c.classid and b.itemid=
            c.itemid and a.userid=d.userid and a.classid=@classid order by newsdate
            desc,newsid desc";
        IList<News>newsList =new List<News>();
        using (SqlConnection connection =new SqlConnection(DBConnection.ConnectString))
        {
            connection.Open();
            SqlCommand command =new SqlCommand(getNewsList, connection);
            command.Parameters.AddWithValue("@classid", classID);
            command.Parameters[0].Value =classID;
            using (SqlDataReader newsReader =command.ExecuteReader())
            {
                while (newsReader.Read())
                {
                    News news =new News();
                    news =FillData(newsReader);
                    newsList.Add(news);
                }
            }
        }
        return newsList;
    }
    /// <summary>
    /// 得到类别页新闻列表
    /// </summary>
    /// <param name="itemID">新闻栏目 ID</param>
    /// <returns>新闻列表</returns>
    public static IList<News>GetClassList(int classID)
    {
        string getNewsList;
        getNewsList ="select a.newsid,a.newstitle,isnull(a.newsource,' ')
            as newsource,isnull(a.newskey,' ') as newskey,a.userid,a.hits,
            a.monthhits,a.classid,a.newsdate,isnull(a.newscontent,'') as newscontent,
            b.itemid,b.itemname,c.classname,d.username from t_news a,t_item b,
            t_class c,t_news_user d where a.classid=c.classid and b.itemid=c.itemid
            and a.userid=d.userid and a.classid=@classid order by newsdate desc,
            newsid desc";
        IList<News>newsList =new List<News>();
        using (SqlConnection connection =new SqlConnection(DBConnection.ConnectString))
        {
            connection.Open();
            SqlCommand command =new SqlCommand(getNewsList, connection);
            command.Parameters.AddWithValue("@classid", classID);
```

```csharp
            using (SqlDataReader newsReader =command.ExecuteReader())
            {
                while (newsReader.Read())
                {
                    News news =new News();
                    news =FillData(newsReader);
                    newsList.Add(news);
                }
            }
        }
        return newsList;
    }
    /// <summary>
    /// 得到单条新闻数据
    /// </summary>
    /// <param name="newsID">新闻 ID</param>
    /// <returns>新闻对象</returns>
    public static News GetSingle(int newsID)
    {
        News news =new News();
        string getNews;
        getNews ="select a.newsid,a.newstitle,isnull(a.newsource,' ') as newsource,isnull(a.newskey,' ') as newskey,a.userid,a.hits, a.monthhits,a.classid,a.newsdate,isnull(a.newscontent,'') as newscontent, b.itemid,b.itemname,c.classname,d.username from t_news a,t_item b, t_class c,t_news_user d where a.classid=c.classid and b.itemid=c.itemid and a.userid=d.userid and a.newsid=@newsid order by newsdate desc,newsid desc";
        using ( SqlConnection connection = new SqlConnection ( DBConnection.ConnectString))
        {
            connection.Open();
            SqlCommand command =new SqlCommand(getNews, connection);
            command.Parameters.AddWithValue("@newsid", newsID);
            using (SqlDataReader newsReader =command.ExecuteReader())
            {
                while (newsReader.Read())
                {
                    news =FillData(newsReader);
                }
            }
        }
        return news;
    }
    /// <summary>
    /// 更新新闻
```

```csharp
/// </summary>
/// <param name="newsTitle">新闻标题</param>
/// <param name="classID">类别ID</param>
/// <param name="newsDate">更新时间</param>
/// <param name="newsKeywords">关键字</param>
/// <param name="newsSource">新闻来源</param>
/// <param name="newsContent">新闻内容</param>
/// <returns>是否成功</returns>
public static bool UpdateNews(string newsTitle, int classID, DateTime newsDate, string newsKeywords, string newsSource, string newsContent, int newsID)
{
    using (SqlConnection connection = new SqlConnection(DBConnection.ConnectString))
    {
        connection.Open();
        SqlCommand command = new SqlCommand("update t_news set newstitle=@newstitle,classid=@classid,newsdate=@newsdate,newskey=@newskey,newsource=@newsource,newscontent=@newscontent where newsid=@newsid", connection);
        command.Parameters.AddWithValue("@newstitle", newsTitle);
        command.Parameters.AddWithValue("@classid", classID);
        command.Parameters.AddWithValue("@newsdate", newsDate);
        command.Parameters.AddWithValue("@newskey", newsKeywords);
        command.Parameters.AddWithValue("@newsource", newsSource);
        command.Parameters.AddWithValue("@newscontent", newsContent);
        command.Parameters.AddWithValue("@newsid", newsID);
        int updateRows = command.ExecuteNonQuery();
        if (updateRows > 0)
        {
            return True;
        }
        else
        {
            return False;
        }
    }
}
/// <summary>
/// 更新文章点击率
/// </summary>
/// <param name="newsID">新闻ID</param>
/// <returns>是否成功</returns>
public static bool UpdateHits(int newsID)
{
```

```csharp
            using (SqlConnection connection = new SqlConnection (DBConnection.
            ConnectString))
            {
                SqlCommand command = new SqlCommand("update t_news set hits=hits+1,
                monthhits=monthhits+1 where newsid=@newsid", connection);
                command.Parameters.Add("@newsid", SqlDbType.Int);
                command.Parameters[0].Value =newsID;
                connection.Open();
                int updateRows = command.ExecuteNonQuery();
                if (updateRows >0)
                {
                    return True;
                }
                else
                {
                    return False;
                }
            }
        }
        /// <summary>
        /// 填充对象数据
        /// </summary>
        /// <param name="newsReader">新闻 Reader</param>
        /// <returns></returns>
        private static News FillData(IDataReader newsReader)
        {
            News news =new News();
            news.NewsID =newsReader.GetInt32(newsReader.GetOrdinal("newsid"));
            news.NewsTitle =newsReader.GetString(newsReader.GetOrdinal("newstitle"));
            news.ClassID =newsReader.GetInt32(newsReader.GetOrdinal("classid"));
            news.ClassName =newsReader.GetString(newsReader.GetOrdinal("classname"));
            news.ItemID =newsReader.GetInt32(newsReader.GetOrdinal("itemid"));
            news.ItemName =newsReader.GetString(newsReader.GetOrdinal("itemname"));
            news.NewsDate =newsReader.GetDateTime(newsReader.GetOrdinal("newsdate"));
            news.NewsKeywords =newsReader.GetString(newsReader.GetOrdinal("newskey"));
            news.NewsSource =newsReader.GetString(newsReader.GetOrdinal("newsource"));
            news.NewsContent =newsReader.GetString(newsReader.GetOrdinal("newscontent"));
            news.UserID =newsReader.GetInt32(newsReader.GetOrdinal("userid"));
            news.UserName =newsReader.GetString(newsReader.GetOrdinal("username"));
            news.Hits =newsReader.GetInt32(newsReader.GetOrdinal("hits"));
            news.MonthHits =newsReader.GetInt32(newsReader.GetOrdinal("monthhits"));
            return news;
        }
    }
}
```

9.4.2 后台代码

后台的代码主要是对数据库中每个表的添加、删除、修改等操作。也就是对新闻的添加编辑，对栏目的添加编辑，对分类的添加编辑，再就是对于用户权限的设置。

下面的代码显示了如何添加新闻和如何编辑新闻。

```
<form id="form1" runat="server">
    <table style="width: 900px;">
        <tr>
            <td class="style1"> </td>
            <td>
                <table style="border: 1px solid #C0C0C0; width:100%;" border="1"
                    cellpadding="1" cellspacing="0" >
                    <tr>
                        <td class="style3">新闻标题：</td>
                        <td class="style2">
                            <asp:TextBox ID="txtNewsTitle" runat="server" Width=
                            "400px"></asp:TextBox>
                            <asp:RequiredFieldValidator ID="rfvNewsTitle" runat=
                            "server" ControlToValidate="txtNewsTitle" Display=
                            "Dynamic" ErrorMessage="新闻标题不能为空！"></asp:
                            RequiredFieldValidator>
                            <asp:RegularExpressionValidator ID="revNewsTitle"
                            runat="server" ControlToValidate="txtNewsTitle"
                            Display="Dynamic" ErrorMessage="禁止输入非法字符！"
                            ValidationExpression="[^%&amp;',;=?$\x22]+"></asp:
                            RegularExpressionValidator>
                        </td>
                    </tr>
                    <tr>
                        <td class="style4">所属类别：</td>
                        <td>
                            <asp:DropDownList ID="dddlNewsClass" runat="server"
                            style="font-size: 12px" AppendDataBoundItems="True">
                            <asp:ListItem Value="None">＝请选择所属类别＝</asp:
                            ListItem>
                            </asp:DropDownList>
                            <asp:RequiredFieldValidator ID="rfvNewsClass" runat=
                            "server" ControlToValidate="dddlNewsClass" Display=
                            "Dynamic" ErrorMessage="新闻类别不能为空！" InitialValue=
                            "None"></asp:RequiredFieldValidator>
                        </td>
                    </tr>
                    <tr>
                        <td class="style4">发布时间：</td>
```

```
        <td>
            <asp:Label ID="lblPublishDate" runat="server">
            </asp:Label>
        </td>
    </tr>
    <tr>
        <td class="style4">关键字：</td>
        <td>
            <asp:TextBox ID="txtKeywords" runat="server" Width="200px"></asp:TextBox>
            <asp:RegularExpressionValidator ID="revKeywords" runat="server" ControlToValidate="txtKeywords" Display="Dynamic" ErrorMessage="禁止输入非法字符！" ValidationExpression="[^%&amp;',;=?$\x22]+"></asp:RegularExpressionValidator>
        </td>
    </tr>
    <tr>
        <td class="style4">新闻来源：</td>
        <td>
            <asp:TextBox ID="txtNewsFrom" runat="server" Width="200px"></asp:TextBox>
            <asp:RegularExpressionValidator ID="revNewsFrom" runat="server" ControlToValidate="txtNewsFrom" Display="Dynamic" ErrorMessage="禁止输入非法字符！" ValidationExpression="[^%&amp;',;=?$\x22]+"></asp:RegularExpressionValidator>
        </td>
    </tr>
    <tr>
        <td class="style4">新闻内容：</td>
        <td>
            <FCKeditorV2:FCKeditor ID="fckNewContent" runat="server" BasePath="FCKeditor/" Height="350px" Width="99%" ToolbarSet="Medium"></FCKeditorV2:FCKeditor>
            <asp:CustomValidator ID="cvNewsContent" runat="server" ErrorMessage="新闻内容不能为空！" ClientValidationFunction="CustomValidate" ControlToValidate="fckNewContent" Display="Dynamic" ValidateEmptyText="True"></asp:CustomValidator>
        </td>
    </tr>
    <tr>
        <td class="style4"> </td>
        <td>
```

```html
                    <asp:Button ID="btnSave" runat="server" Text="保存"
                    onclick="btnSave_Click" />
                      <input id="btnReset" type="Reset" value="重
                    置" /></td>
                </tr>
            </table>
        </td>
    </tr>
</table>
</form>
```

对应的.cs文件如下：

```csharp
using NewsManage.Model;
using NewsManage.DAL;
public partial class Admin_AddNews : System.Web.UI.Page
{
    Administrator admin =new Administrator();
    protected void Page_Load(object sender, EventArgs e)
    {
        if (Session["CurrentUser"] ==null)
        {
            Page.RegisterStartupScript("err1", "<script>parent.window.opener=
            'anyone';window.open('Loggin.aspx');parent.window.close();</script>");
        }
        else
        {
            admin = (Administrator)Session["CurrentUser"];
            lblPublishDate.Text = string.Format("{0:yyyy年MM月dd日}", DateTime.
            Today);
            if (!IsPostBack)
            {
                dddlNewsClass.DataTextField ="ClassName";
                dddlNewsClass.DataValueField ="ClassID";
                if (admin.Power)
                {
                    IList<NewsClass>newsClassListAll =new List<NewsClass>();
                    newsClassListAll =NewsClassDB.GetAll();
                    dddlNewsClass.DataSource =newsClassListAll;
                    dddlNewsClass.DataBind();
                }
                else
                {
                    IList<NewsClass>newsClassList =new List<NewsClass>();
                    newsClassList =NewsClassDB.GetListByUser(admin.UserID);
                    dddlNewsClass.DataSource =newsClassList;
```

```csharp
                dddlNewsClass.DataBind();
            }
        }
    }
    protected void btnSave_Click(object sender, EventArgs e)
    {
        string newsTitle,newsKeywords,newsSource,newsContent;
        newsTitle =txtNewsTitle.Text;
        if (txtKeywords.Text.Trim() !="")
        {
            newsKeywords =txtKeywords.Text.Trim();
        }
        else
        {
            newsKeywords =" ";
        }
        if (txtNewsFrom.Text.Trim() !="")
        {
            newsSource =txtNewsFrom.Text.Trim();
        }
        else
        {
            newsSource =" ";
        }
        newsContent =HttpUtility.HtmlEncode(fckNewContent.Value);

        if ( NewsDB. AddNews ( newsTitle, Convert. ToInt32 ( dddlNewsClass. SelectedValue), DateTime. Today, newsKeywords, newsSource, newsContent, admin.UserID))
        {
            txtNewsTitle.Text ="";
            txtKeywords.Text ="";
            txtNewsFrom.Text ="";
            fckNewContent.Value ="";
            Page.RegisterStartupScript("ok1", "<script>alert('添加新闻成功!');</script>");
        }
        else
        {
            Page.RegisterStartupScript("err1", "<script>alert('添加失败,请稍后重试!')</script>");
        }
    }
}
```

对于新闻的编辑如何实现,代码如下:

```
<form id="form1" runat="server">
    <table style="width: 790px;">
        <tr>
            <td>
                  检索关键字:<asp:TextBox ID="txtNewsKeywords" runat=
                "server" Width="230px"></asp:TextBox>
                  <asp:DropDownList ID="dddlNewsClass" runat="server"
                AppendDataBoundItems="True" style="font-size: 12px">
                    <asp:ListItem Value="%">=请选择新闻类别=</asp:ListItem>
                </asp:DropDownList>
                  <asp:Button ID="btnRetrieve" runat="server" Text=
                "检索新闻" />
            </td>
        </tr>
        <tr>
            <td>
                <asp:GridView ID="gdvNews" runat="server" BackColor="White"
                BorderColor="#3366CC" BorderStyle="None" BorderWidth="1px"
                CellPadding="4"
                DataSourceID="odsNews" AllowPaging="True" AutoGenerateColumns=
                "False" PageSize="15" Width="780px" ondatabound="gdvNews_DataBound"
                DataKeyNames="NewsID">
                <PagerSettings FirstPageText="首页" LastPageText="末页" Mode=
                "NumericFirstLast" NextPageText="下一页" PreviousPageText="上一页" />
                <FooterStyle BackColor="#99CCCC" ForeColor="#003399" />
                <RowStyle BackColor="White" ForeColor="#003399" />
                <Columns>
                    <asp:BoundField DataField="NewsID" HeaderText="编号"
                    SortExpression="NewsID">
                        <ItemStyle Width="50px" />
                    </asp:BoundField>
                    <asp:BoundField DataField="NewsTitle" HeaderText="新闻标题"
                        SortExpression="NewsTitle">
                        <ItemStyle Width="280px" />
                    </asp:BoundField>
                    <asp:BoundField DataField="ClassName" HeaderText="所属类别"
                        SortExpression="ClassName">
                        <ItemStyle Width="100px" />
                    </asp:BoundField>
                    <asp:BoundField DataField="NewsDate" DataFormatString=
                    "{0:yyyy-MM-dd}"
                        HeaderText="发布时间" SortExpression="NewsDate">
                        <ItemStyle Width="80px" />
                    </asp:BoundField>
```

```
        <asp:BoundField DataField="NewsSource" HeaderText="新闻来源"
            SortExpression="NewsSource" HtmlEncode="False">
            <ItemStyle Width="100px" />
        </asp:BoundField>
        <asp:BoundField DataField="Hits" HeaderText="点击"
            SortExpression="Hits">
            <ItemStyle Width="50px" />
        </asp:BoundField>
        <asp:HyperLinkField DataNavigateUrlFields="newsid"
            DataNavigateUrlFormatString="EditNews.aspx?NewsID={0}"
            HeaderText="编辑"
            Text="编辑">
            <ItemStyle Width="30px" />
        </asp:HyperLinkField>
        <asp:TemplateField HeaderText="删除" ShowHeader="False">
            <ItemTemplate>
                <asp:LinkButton ID="lbDelete" runat="server" CausesValidation="False" CommandName="Delete" Text="删除" ></asp:LinkButton>
            </ItemTemplate>
            <ItemStyle Width="30px" />
        </asp:TemplateField>
    </Columns>
    <PagerStyle BackColor="#CCCCCC" ForeColor="#003399" HorizontalAlign="Right" />
    <SelectedRowStyle BackColor="#009999" Font-Bold="True" ForeColor="#CCFF99" />
    <HeaderStyle BackColor="#003399" Font-Bold="True" ForeColor="#CCCCFF" />
</asp:GridView>
<asp:ObjectDataSource ID="odsNews" runat="server" SelectMethod="GetList" TypeName="NewsManage.DAL.NewsDB" DeleteMethod="DeleteNews">
    <SelectParameters>
        <asp:ControlParameter ControlID="dddlNewsClass" DefaultValue="%" Name="classID" PropertyName="SelectedValue" Type="String" />
        <asp:ControlParameter ControlID="txtNewsKeywords" DefaultValue="%" Name=newsKeywords" PropertyName="Text" Type="String" />
    </SelectParameters>
</asp:ObjectDataSource>
            </td>
        </tr>
</table>
<br />
<asp:HyperLink ID="hlAddNews" runat="server" NavigateUrl="~/Admin/AddNews.aspx">添加新闻</asp:HyperLink>
```

```
</form>
```

对应的.cs文件如下：

```csharp
using NewsManage.DAL;
using NewsManage.Model;
public partial class Admin_ManageNews : System.Web.UI.Page
{
    Administrator admin =new Administrator();
    protected void Page_Load(object sender, EventArgs e)
    {
        if (Session["CurrentUser"] ==null)
        {
            Page.RegisterStartupScript("err1", "<script>parent.window.opener=
            'anyone';window.open('Loggin.aspx');parent.window.close();</script>");
        }
        else
        {
            admin = (Administrator)Session["CurrentUser"];
            if (!IsPostBack)
            {
                dddlNewsClass.DataTextField ="ClassName";
                dddlNewsClass.DataValueField ="ClassID";
                if (admin.Power)
                {
                    IList<NewsClass>newsClassListAll =new List<NewsClass>();
                    newsClassListAll =NewsClassDB.GetAll();
                    dddlNewsClass.DataSource =newsClassListAll;
                    dddlNewsClass.DataBind();
                }
                else
                {
                    IList<NewsClass>newsClassList =new List<NewsClass>();
                    newsClassList =NewsClassDB.GetListByUser(admin.UserID);
                    dddlNewsClass.DataSource =newsClassList;
                    dddlNewsClass.DataBind();
                }
            }
        }
    }
    protected void gdvNews_DataBound(object sender, EventArgs e)
    {
        for (int i =0; i <gdvNews.Rows.Count; i++)
        {
            gdvNews.Rows[i].Attributes.Add("onmouseover", "c = this.style.
            backgroundColor;this.style.backgroundColor='#eeeeee'");
```

```
            gdvNews.Rows[i].Attributes.Add(" onmouseout ", " this.style.
            backgroundColor=c");
            LinkButton deleteButton =new LinkButton();
            deleteButton = (LinkButton)gdvNews.Rows[i].Cells[7].FindControl
            ("lbDelete");
            deleteButton.Attributes.Add("onclick", "return confirm('确定要删除此
            新闻吗?')");
        }
    }
}
```

上面的代码中使用了 NewsClassDB 这个类，下面把这个类中的方法展示，在本系统中还有好多类的定义，这里就不一一列出，请读者参阅源代码。

```
using System.Data.SqlClient;
using NewsManage.Model;
namespace NewsManage.DAL
{   /// <summary>
    ///NewsClassDB 的摘要说明
    /// </summary>
    public class NewsClassDB
    {   /// <summary>
        /// 新增类别数据
        /// </summary>
        /// <param name="className">类别名称</param>
        /// <param name="classDescription">类别描述</param>
        /// <param name="classOrder">类别顺序</param>
        /// <param name="itemID">所属栏目 ID</param>
        /// <returns>是否成功</returns>
        public static bool AddClass(string className, string classDescription, int
        classOrder, int itemID)
        {
            using (SqlConnection connection = new SqlConnection(DBConnection.
            ConnectString))
            {
                SqlCommand command =new SqlCommand("insert into t_class values
                (@classname,@classdesc,@classorder,@itemid)", connection);
                connection.Open();
                command.Parameters.Add("@classname", SqlDbType.VarChar, 50);
                command.Parameters.Add("@classdesc", SqlDbType.VarChar, 200);
                command.Parameters.Add("@classorder", SqlDbType.Int);
                command.Parameters.Add("@itemid", SqlDbType.Int);
                command.Parameters[0].Value =className;
                command.Parameters[1].Value =classDescription;
                command.Parameters[2].Value =classOrder;
```

```
            command.Parameters[3].Value =itemID;
            int effectiveRows =command.ExecuteNonQuery();
            if (effectiveRows >0)
            {
                return True;
            }
            else
            {
                return False;
            }
        }
    }
    /// <summary>
    /// 得到所有类别
    /// </summary>
    /// <returns>类别列表</returns>
    public static IList<NewsClass>GetAll()
    {
        IList<NewsClass>newsClassList =new List<NewsClass>();
        using (SqlConnection connection = new SqlConnection(DBConnection.
        ConnectString))
        {
            SqlCommand command =new SqlCommand("select * from t_class order by
            itemid,classorder,classid", connection);
            connection.Open();
            SqlDataReader classReader =command.ExecuteReader();
            while (classReader.Read())
            {
                NewsClass newsClass =new NewsClass();
                newsClass =FillData(classReader);
                newsClassList.Add(newsClass);
            }
        }
        return newsClassList;
    }
    /// <summary>
    /// 得到栏目下所有类别
    /// </summary>
    /// <param name="itemID">栏目 ID</param>
    /// <returns>类别列表</returns>
    public static IList<NewsClass>GetItemClass(int itemID)
    {
        IList<NewsClass>newsClassList =new List<NewsClass>();
        using (SqlConnection connection = new SqlConnection(DBConnection.
```

```csharp
            ConnectString))
        {
            SqlCommand command = new SqlCommand("select * from t_class where
            itemid=@itemid order by itemid,classorder,classid", connection);
            connection.Open();
            command.Parameters.Add("@itemid", SqlDbType.Int);
            command.Parameters[0].Value = itemID;
            SqlDataReader classReader = command.ExecuteReader();
            while (classReader.Read())
            {
                NewsClass newsClass = new NewsClass();
                newsClass = FillData(classReader);
                newsClassList.Add(newsClass);
            }
        }
        return newsClassList;
}
/// <summary>
/// 得到所有的类别以及所对应的栏目名称
/// </summary>
/// <returns>类别栏目对象集合</returns>
public static IList<NewsClassItem> GetClassItemName()
{
    IList<NewsClassItem> newsClassItemList = new List<NewsClassItem>();
    using (SqlConnection connection = new SqlConnection(DBConnection.
    ConnectString))
    {
        SqlCommand command = new SqlCommand("select a.classid,a.classname,a.
        classdesc,a.classorder,b.itemid,b.itemname,b.itemdesc,b.itemorder
        from t_Class a,t_Item b where a.itemid=b.itemid order by classid",
        connection);
        connection.Open();
        using (SqlDataReader classItemReader = command.ExecuteReader())
        {
            while (classItemReader.Read())
            {
                NewsClassItem newsClassItem = new NewsClassItem();
                newsClassItem.ClassID = classItemReader.GetInt32
                (classItemReader.GetOrdinal("classid"));
                newsClassItem.ClassName = classItemReader.GetString
                (classItemReader.GetOrdinal("classname"));
                newsClassItem.ClassDescription = classItemReader.GetString
                (classItemReader.GetOrdinal("classdesc"));
                newsClassItem.ClassOrder = classItemReader.GetInt32
```

```csharp
                    (classItemReader.GetOrdinal("classorder"));
                newsClassItem.ItemID = classItemReader.GetInt32
                    (classItemReader.GetOrdinal("itemid"));
                newsClassItem.ItemName = classItemReader.GetString
                    (classItemReader.GetOrdinal("itemname"));
                newsClassItem.ItemDescription = classItemReader.GetString
                    (classItemReader.GetOrdinal("itemdesc"));
                newsClassItem.ItemOrder = classItemReader.GetInt32
                    (classItemReader.GetOrdinal("itemorder"));
                newsClassItemList.Add(newsClassItem);
            }
        }
        return newsClassItemList;
    }
}
/// <summary>
/// 得到管理员所管辖的类别列表
/// </summary>
///
/// <returns>类别列表</returns>
public static IList<NewsClass> GetListByUser(int userID)
{
    IList<NewsClass> newsClassList = new List<NewsClass>();
    using (SqlConnection connection = new SqlConnection(DBConnection.ConnectString))
    {
        SqlCommand command = new SqlCommand("select a.* from t_class a,t_popedom b where a.classid=b.classid and b.userid=@userid order by itemid,classorder,classid", connection);
        connection.Open();
        command.Parameters.Add("@userid", SqlDbType.Int);
        command.Parameters[0].Value = userID;
        SqlDataReader classReader = command.ExecuteReader();
        while (classReader.Read())
        {
            NewsClass newsClass = new NewsClass();
            newsClass = FillData(classReader);
            newsClassList.Add(newsClass);
        }
    }
    return newsClassList;
}
/// <summary>
/// 删除类别
```

```csharp
/// </summary>
/// <param name="classID">类别 ID</param>
public static void DeleteClass(int classID)
{
    using (SqlConnection connection = new SqlConnection(DBConnection.ConnectString))
    {
        connection.Open();
        SqlCommand commandSelect = new SqlCommand("select count(*) from t_news where classid=@classid", connection);
        commandSelect.Parameters.Add("@classid", SqlDbType.Int);
        commandSelect.Parameters[0].Value = classID;
        SqlCommand commandDelete = new SqlCommand("delete from t_class where classid=@classid", connection);
        commandDelete.Parameters.Add("@classid", SqlDbType.Int);
        commandDelete.Parameters[0].Value = classID;

        int selectRows = Convert.ToInt32(commandSelect.ExecuteScalar());
        if (selectRows > 0)
        {
            HttpContext.Current.Response.Write("<script>alert('该类别下有文章数据,不能删除!');location.replace(window.location.href);</script>");
        }
        else
        {
            int deleteRows = commandDelete.ExecuteNonQuery();
            if (deleteRows > 0)
            {
                //HttpContext.Current.Response.Write("<script>alert('删除成功!');</script>");
            }
            else
            {
                HttpContext.Current.Response.Write("<script>alert('删除失败,请稍后重试!');</script>");
            }
        }
    }
}
/// <summary>
/// 更新类别
/// </summary>
/// <param name="classID">类别 ID</param>
/// <param name="className">类别名称</param>
```

```csharp
/// <param name="classDescription">类别描述</param>
/// <param name="classOrder">类别顺序</param>
/// <param name="itemID">所属栏目ID</param>
public static void UpdateClass(int classID, string className, string classDescription, int classOrder, int itemID)
{
    using (SqlConnection connection = new SqlConnection(DBConnection.ConnectString))
    {
        SqlCommand commandUpdate = new SqlCommand("update t_class set classname=@classname, classdesc=@classdesc, classorder=@classorder, itemid=@itemid where classid=@classid", connection);
        connection.Open();
        commandUpdate.Parameters.Add("@classid", SqlDbType.Int);
        commandUpdate.Parameters.Add("@classname", SqlDbType.VarChar, 50);
        commandUpdate.Parameters.Add("@classdesc", SqlDbType.VarChar, 200);
        commandUpdate.Parameters.Add("@classorder", SqlDbType.Int);
        commandUpdate.Parameters.Add("@itemid", SqlDbType.Int);
        commandUpdate.Parameters[0].Value = classID;
        commandUpdate.Parameters[1].Value = className;
        commandUpdate.Parameters[2].Value = classDescription;
        commandUpdate.Parameters[3].Value = classOrder;
        commandUpdate.Parameters[4].Value = itemID;

        int updateRows = commandUpdate.ExecuteNonQuery();
        if (updateRows > 0)
        {
            HttpContext.Current.Response.Write("<script>alert('更新成功!');</script>");
        }
        else
        {
            HttpContext.Current.Response.Write("<script>alert('更新失败,请稍后重试!');</script>");
        }
    }
}
private static NewsClass FillData(IDataReader classReader)
{
    NewsClass newsClass = new NewsClass();
    newsClass.ClassID = classReader.GetInt32(classReader.GetOrdinal("classid"));
    newsClass.ClassName = classReader.GetString(classReader.GetOrdinal("classname"));
```

```csharp
            newsClass.ClassDescription = classReader.GetString(classReader.GetOrdinal
            ("classdesc"));
            newsClass.ClassOrder = classReader.GetInt32(classReader.GetOrdinal
            ("classorder"));
            newsClass.ItemID = classReader.GetInt32(classReader.GetOrdinal
            ("itemid"));
            return newsClass;
        }
    }
}
```

参 考 文 献

[1] 朱玉超,鞠艳,王代勇.ASP.NET 项目开发教程[M].北京：电子工业出版社,2008.
[2] 刘乃丽.完全手册 ASP.NET 2.0 网路开发详解[M].北京：电子工业出版社,2008.
[3] 刘丹妮.ASP.NET 2.0(C♯)大学实用教程[M].北京：电子工业出版社,2009.
[4] 沈士根,汪承焱,许小东.Web 程序设计：ASP.NET 实用网站开发[M].北京：清华大学出版社,2009.
[5] 邓文渊.ASP.NET 2.0(C♯)大学实用教程[M].北京：机械工业出版社,2009.
[6] 龚赤兵.ASP.NET 2.0 网站开发案例教程[M].北京：中国水利水电出版社,2009.